# The Economics of Non-Market Goods and Resources

Volume 16

**Series Editor**

Ian J Bateman, ENV, CSERGE, School University of East Anglia, Exeter, UK

More information about this series at http://www.springer.com/series/5919

Anthony C. Fisher

# Lecture Notes on Resource and Environmental Economics

To Dick – In friendship,
Anthony C. Fisher

Springer

Anthony C. Fisher
Department of Agricultural
and Resource Economics
University of California
Berkeley, CA, USA

ISSN 1571-487X
The Economics of Non-Market Goods and Resources
ISBN 978-3-030-48957-1 ISBN 978-3-030-48958-8 (eBook)
https://doi.org/10.1007/978-3-030-48958-8



This Springer imprint is published by the registered company Springer Nature Switzerland AG
The registered company address is: Gewerbestrasse 11, 6330 Cham, Switzerland

# Preface

This volume is based on lectures, primarily in classes, both graduate and undergraduate, plus seminar and conference presentations, that I have given over the past 40+ years, 40 at UC Berkeley, in the field of natural resource and environmental economics. Although some of the material is similar to what one would find in a textbook and is indeed adapted from one or another, as well as from journal articles by myself and others, it represents my take on a subject—what has been of particular interest to me or perhaps what I was learning, working through, and lecturing about at a given time. This has, not surprisingly, also resulted in changes or additions to the original source, indicated as appropriate. There is no attempt at balance in coverage across subjects or length of treatment, as one would expect to find in a standard text. A couple of examples are given below to illustrate what I mean and to provide an explanation.

In the summers of 1975 and 1976, in the wake of the "energy crisis," triggered by the approximately 300 percent jump in world oil prices in 1973–74, along with the OPEC oil embargo on the U.S, I attended workshops organized by Bob Solow and Joe Stiglitz intended for theorists to focus on the newly important economics of natural resources, with an emphasis on nonrenewable or exhaustible resources such as oil. Not surprisingly, I learned a good deal and spent the next several years working through some of this material and presenting it in class lectures, along with the occasional paper. This is one reason why the notes on exhaustible resources are given so much more space than those on renewable resources such as timber. A second reason for the imbalance is that, once one has a model for optimal depletion of exhaustible resources, it can be adapted to describe the optimal use of renewables by adding a biological growth or regeneration function. Of course different issues are raised, especially with respect to particular resources, and some of these, such as the desirability of maximum sustained yield and prospects for exhaustion, are discussed in the notes on renewable resources. This section also features an in-depth treatment of timber harvesting, which as I shall suggest not only fits nicely with a generic treatment of renewable resources but leads quite naturally to the following section, notes on environmental resources.

Here I need to mention a technical issue: the mathematics used in these two sections. The results in the first, on optimal depletion of exhaustible resources, are generated in two ways: using only elementary calculus, and then optimal control, a method of dynamic optimization, to yield additional insights into the problem. This is preceded by an introduction to the method, so the reader will be able to apply it to the depletion problem, as in the text. But again, the main results are obtained in the first approach as well, and in both cases, I emphasize the economic common sense or interpretation. Here as elsewhere the approach is intuitive, rather than rigorously mathematical. In the notes on optimal use of renewable resources, results are again obtained in two ways, first using elementary calculus as applied to timber harvesting, and then by optimal control, extending the model developed for the depletion problem. In both sections, notes on the method and application of optimal control are starred and can be skipped if desired.

The notes on environmental resources are also idiosyncratic, focusing on the economics of natural environments, which can be regarded simply as sites from which commercial natural resources such as timber or minerals can be extracted or as yielding benefits in a natural or preserved state as well, implying a need to take these into account in a benefit/cost analysis of a project to develop the environment for commercial resource extraction. What is neglected in this volume, with one exception discussed below, is the main topic in the literature on environmental economics, the optimal control of pollution, including both choice of method—direct controls vs. market-like instruments such as a pollution tax or cap-and-trade, and empirical analyses of the costs and benefits of types and levels of control in particular cases.

My interest in the economics of natural environments dates from the early 1970s at Resources for the Future (RFF), where I came to work with John Krutilla, whose path-breaking 1967 paper, Conservation Reconsidered, in the American Economic Review, first showed how economics could play a role in policy and management decisions about the disposition of such areas. With John and others, at RFF and subsequently elsewhere, I worked on both the theory of optimal use under conditions of uncertainty and irreversibility and empirical studies of particular cases. Both conditions – uncertainty and irreversibility—are crucial here, as once a natural environment is transformed, restoration will not be possible, so an assessment of costs and benefits needs to go far into the future, with uncertainty increasing along the way. This in turn leads to the concept of option value, which I discuss with the aid of a computation and an illustrative application. Full credit is given to co-authors for material originally appearing elsewhere.

At around the same time, early to middle 1970s, I became interested in a larger, more general problem: is modern civilization using resources, including the assimilative capacity of the environment, at a rate which cannot be sustained? This interest was stimulated by the energy and environmental "crises" of the time and also by the publication of the widely read and influential volume, "The Limits to Growth," which presents computer simulations of the dynamic interactions among traditional macroeconomic variables such as labor (population), capital, and output, plus agriculture, natural resources, and the environment. The striking result of the simulations, based on a variety of different assumptions, for example about stocks of resources including agricultural land, is that in the absence of a number of drastic changes in policies affecting population and nature of the economy, resource and environmental limits will prevent expansion of the economy to meet the needs of a growing population, and will in fact lead to a collapse of the economy and the population it supports. The authors are not economists and the methods and results have been criticized by economists, as I discuss in the section on resources, growth, and sustainability.

Lectures in this section also consider in some detail alternative measures of resource scarcity, including reserves, reserves/production ratios, costs, prices, and royalties. The question is: How do we know if we are running out of resources? The popular approach is to look only at physical measures such as those involving reserves, but as I discuss here this is too simple and almost certainly misleading without attention to the economic concepts. Theoretical properties of these concepts are considered, and empirical results drawn from diverse sources presented and discussed.

In a sense this is a very old concern in economics, going back at least to the first (1798) edition of Robert Malthus's Essay on Population, which dealt with the relationship between a rapidly growing population and at best a more slowly growing agricultural land resource. Circling back, this is the subject of the first set of lectures—what I call the "classical roots of resource economics"—following a brief introduction offering some thoughts on why we study resource (and environmental) economics today. My discussion of the "classical roots" is based on a more extensive treatment in another iconic RFF work, "Scarcity and Growth," by Harold Barnett and Chandler Morse, published in 1963 and focusing on what I argue are the still-relevant ideas of Malthus, David Ricardo, and John Stuart Mill.

Going forward, the modern formulation of the problem falls under the heading of "sustainability." Though one can no doubt find earlier references, the concept of sustainability was brought to the attention of the general public, policy analysts and economists by the 1987 report of the World Commission on Environment and Development, where the term was defined as: "Development that meets the needs of the present without compromising the ability of future generations to meet their own needs." The task of economists has been to translate this very broad and rather vague statement into a form suitable for measurement, at least in theory. This set of lectures takes up the task, first in an intuitive development of the analytics, then in a somewhat more rigorous formulation, and finally with some fragmentary but illustrative empirical results.

This brings us to the final section, on the economics of climate change, a major focus of my research, teaching, and lecturing over the past two decades. I hasten to add that my interests, where mentioned in the notes, represent only a small fraction of the new field, which has grown dramatically in recent years, perhaps as a result of dramatic events—droughts, floods, wildfires, and intense tropical storms—predicted impacts of climate change, though of course some will have occurred, perhaps less violently, in a stable climate regime. This section is intended not so much as a contribution to one or another strand of the literature, rather as an attempt to bridge what I see as the disconnect between climate science and climate economics. As I argue, it seems fair to say that climate scientists (and many other natural scientists) believe that climate change is a major problem, perhaps the most important one facing us this century, whereas at least until very recently, much of the established economic literature seems to suggest that global warming is not an issue that requires dramatic action in the near term (the impacts will be minor, perhaps even beneficial, other issues are more important, and so on).

The first part of the section presents in some detail relevant (and alarming) findings concerning already existing changes, prospects for future changes, and potential impacts toward the end of the century and beyond. Most of this information is vintage 2014, when the paper on which it is based was written, but my impression is that it is confirmed by more recent findings, for example, on the more rapid rise in sea level over the past couple of decades. I then turn to what I believe are the most relevant economic concepts here: discounting for the long run, and irreversibilities, both introduced in earlier sections and in my judgment specially relevant in the context of climate change. Even if one accepts this, the question remains: what are the implications for policy, in a world beset by other potentially catastrophic events such as detonation of a nuclear weapon in a major city, spread of a mega-virus, and so on, in addition to ongoing problems, environmental, and otherwise. My thoughts on this question, followed by a brief discussion of policy instruments, close the section.

## Disclaimers

I have said this is not a text, because of the uneven and idiosyncratic coverage of topics. But it could be useful as a supplement to a text, precisely because of this feature. For example, if students or an instructor would like to learn more about the economics of natural environments, not covered in standard texts, certainly not in the detail presented here—theory of optimal use, valuation of alternative uses, option value, an illustrative application—the material presented here will be useful.

This is also not a review of the literature on environmental and resource economics, necessarily because of the selective coverage but also because for a class lecture or even a conference presentation, there is no obligation or even an ability to reference everything written on a subject—although as it has turned out, the list of references is fairly extensive. If I may speak to my colleagues here, please understand that if one of your papers is not referenced, it is not because I think it unimportant, rather that it has been necessary to select just a few which have specially influenced my work or which will be accessible to the intended audience, such as undergraduate students. On the other hand, there may be references here, for example, from the scientific literature on climate change, which will be unfamiliar to economists working in the area.

Finally, this is not a research monograph or even a presentation of results of new research. Of course results of research, my own and others, are presented and discussed, but apart from citation of a few new empirical results, a couple of new wrinkles on the analysis of optimal use of renewable resources, and computational analyses of optimal use of both renewable and exhaustible resources, new research is not a strength of the volume.

## Acknowledgments

Because it is written in a foreign language with which I am not familiar—LaTeX—this book could not have been done without the participation and assistance of two recent graduates of the Ph.D. program in Agricultural and Resource Economics at UC Berkeley: Leslie Martin, who set up the LaTeX framework and contributed the subsection on discounting in the lecture notes on exhaustible resources, and a diagrammatic exposition of the distinction between Ricardian rents and Hotelling royalties; and Daniel Tregeagle, for many contributions throughout the volume, especially for the computational analyses of the optimal use of exhaustible and renewable resources, for the development and presentation of material on empirical measures of resource scarcity, for providing a more professional presentation of tables and figures there and elsewhere, and for formatting generally.

In addition, I want to acknowledge indirect contributions of two other former students, graduates of the Ph.D. programs in (respectively) Agricultural and Resource Economics, and Economics: Phu Viet Le, who provided research assistance and co-authored the paper on climate science and climate economics, from which the lecture notes on the subject are drawn; and (Heidi) Jo Albers, who, along with Michael Hanemann, co-authored the paper on the valuation and management of tropical forests under uncertainty, from which the lecture notes on that subject are drawn.

I have indicated that co-authors are of course credited where discussion in the text is based on joint work. But I would like to single out in particular Michael Hanemann, my colleague at UC Berkeley, for many stimulating discussions over the years, which have in turn led to publications in the areas of decisions under uncertainty and irreversibility, the definition and computation of option value, and aspects of the economics of climate change.

Berkeley, USA Anthony C. Fisher

# Contents

# Chapter 1
# The Classical Roots of Resource Economics

**Abstract** This chapter begins by asking why we study resource and environmental economics, and answers: To know what economic theory and related empirical findings can tell us about a variety of interesting and important issues in this area, from rates of use of extractive resources, to more recent concerns about sustainability and climate change. All can be considered to derive from the Classical Economists, Malthus and Ricardo, who focused on the most important resource of the time, agricultural land. For Malthus, population (labor) was growing exponentially while land was fixed in supply, or growing very slowly if at all, leading ultimately by the law of diminishing returns to a large population living at the subsistence level. Ricardo shifted the focus to points along the way, where land of a given quality was fixed, but cultivation could shift to the next best land when diminishing returns set in, giving rise to the phenomenon of rent, or the return to each unit of the higher quality land. An example of the computation of rent is given, and the chapter closes with a short discussion of the insights and deficiencies of the classical analysis.

## 1.1 Introduction: Why Do We Study Resource Economics?

We study environmental and resource economics because we want to know what economic theory and related empirical findings can tell us about a variety of interesting and important issues in this area. Some current examples include sustainability, loss of biodiversity, and global climate change. Some earlier examples include the conservation movement (1890–1920), focusing on stocks of commercial extractive resources such as wood, iron, and coal, and the echo in the late 1940s and early 1950s, when a Presidential Commission was set up to forecast emerging extractive resource shortages in the wake of heavy depletion during World War II; the "energy crisis" of the early 1970s, repeated in the late 1970s and early 1980s, and on occasion since, triggered by oil price shocks or supply disruptions; and the "environmental crisis" of the early 1970s and on occasion since, leading to air and water quality legislation and regulation and the founding of the Environmental Protection Agency (EPA).

What is the common thread in these examples? We study environmental and resource economics because we are concerned about running out of resources,



A. C. Fisher, *Lecture Notes on Resource and Environmental Economics*, The Economics of Non-Market Goods and Resources 16,
https://doi.org/10.1007/978-3-030-48958-8_1

whether the conventional extractive resources of the early conservation movement, or the capacity of environmental media such as the oceans and the atmosphere for receiving and assimilating the residuals of economic activity.

Given that resources are scarce, and perhaps growing scarcer, we want to know, how do we make the best use of them? And to what extent is optimal use accomplished by markets? But it's not just resource economics. Why do we study economics? Again, because we are concerned about scarcity. A common definition of economics is: the study of the allocation of scarce resources among various and competing ends. And what is scarce? According to the classical economists such as Robert Malthus and David Ricardo, land—mainly agricultural land, but also minerals on the land. So the larger discipline of economics, not just resource and environmental economics, has its roots in a concern for the adequacy of the natural resource base to sustain living standards. The measurement of resource scarcity is an empirical problem, though as we shall suggest, guided by theory. And theory is of course central in answering questions about optimal use of resources and the extent to which this is accomplished by markets.

## 1.2 The Classical Economists: Robert Malthus

According to Malthus's first Essay on Population (Malthus 1798), labor is population, and population, unchecked, grows "geometrically", or what we would now call exponentially, at a constant percentage rate per unit of time. This is the phenomenon of compound interest, which as we know can lead to very large numbers.

The difficulty is that land, the other resource input for Malthus, grows only "arithmetically", i.e., by small increments, if at all. This in turn implies the famous "law of diminishing returns", or in other words, beyond some point, the marginal product of the variable factor, labor, applied to the fixed factor, land, must be diminishing.

Malthus spelled out several scenarios. Since population growth must be checked, there are two possibilities: the "preventive" checks, in other words, forestalling additional growth, and the "positive" checks, or methods of reducing the existing population. The preventive checks are "moral restraint" and "vice".

What did Malthus mean by these somewhat archaic terms? Moral restraint is simply refraining from sexual activity; the meaning of vice is less clear but appears to include contraception, especially in the context of prostitution, and homosexuality. Since Malthus did not expect much in the way of moral restraint, given his understanding of human nature, and vice he as a clergyman did not accept, that leaves only the positive checks: "famine", "pestilence", and "war", all varying forms of "misery", hence the subsequent characterization of economics as "the dismal science". This is expressed in what Kenneth Boulding has called the *Dismal Theorem*: If the only check on population growth is misery, the population will grow until it is miserable. The Malthusian scenario can be represented in a simple diagram, as in Fig. 1.1, showing diminishing returns beyond point Q leading to point R, the subsistence level, at which the positive checks operate.

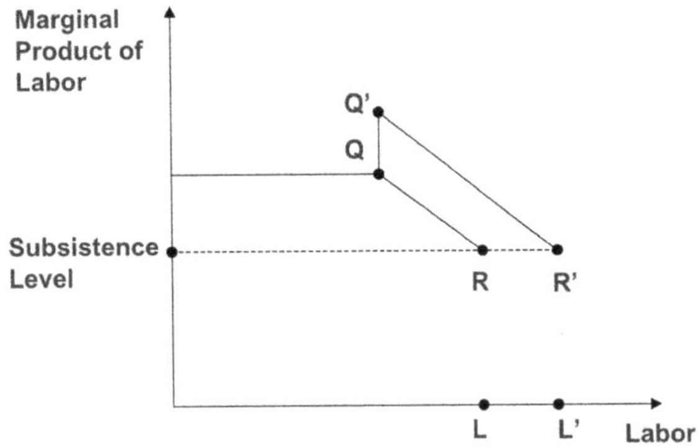


**Fig. 1.1** Malthusian scenario: diminishing marginal returns pushes the marginal product of labor back to subsistence levels

In fact, as shown on Fig. 1.1, the economic prospect is even worse than the Dismal Theorem suggests. Consider an improvement in the arts of cultivation that raises the productivity of labor, as in the move from Q to Q'. The inexorable law of diminishing returns sets in, with the result that the population again winds up at the subsistence level, R', as stated in the *Utterly Dismal Theorem*: Any improvement which raises labor productivity and income will only result in a greater number of people living in misery.

## 1.3 The Classical Economists: David Ricardo

Ricardo's Principles of Political Economy was first published in 1817 (Ricardo 1817). Ricardo is both more and less pessimistic than Malthus. Whereas Malthus envisioned a fixed quantity of arable land, perhaps "arithmetically" enlarged by draining swamps and the like, beyond which diminishing returns to population growth would lead to the subsistence level, Ricardo doesn't speak of any such ultimate or fixed resource constraint. On the other hand, he argues that diminishing returns in fact set in long before the last unit of land is brought into cultivation. Rather, land is what we would call a heterogeneous input to production, that is, it varies in quality. After all of the best land is in production, further growth in population or labor leads to diminishing returns, or equivalently, in the Ricardian analysis, rising costs of producing additional output, as shown in Fig. 1.2.

Beyond some point, where the costs of production rise to the level of those from the next best land, any additional labor is applied to that land—and so on, through as many different types or qualities of land as it may be useful to distinguish. The key assumption, or feature of this model, is that each type or quality of land is fixed, in limited supply, so that as the economy grows and the demand for food or fiber increases, increasingly poorer qualities of land must be brought into production.

Ricardo also broadened the focus of the analysis in a way that is particularly relevant to us: he included resources on or under the land, minerals, speaking of

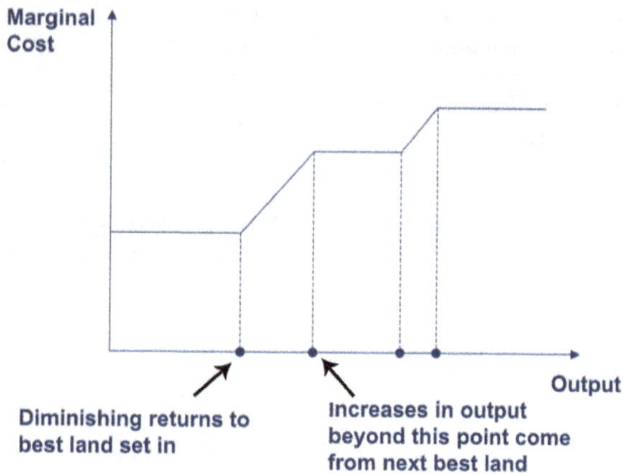


**Fig. 1.2** Ricardian model: land is used in order of increasing marginal cost

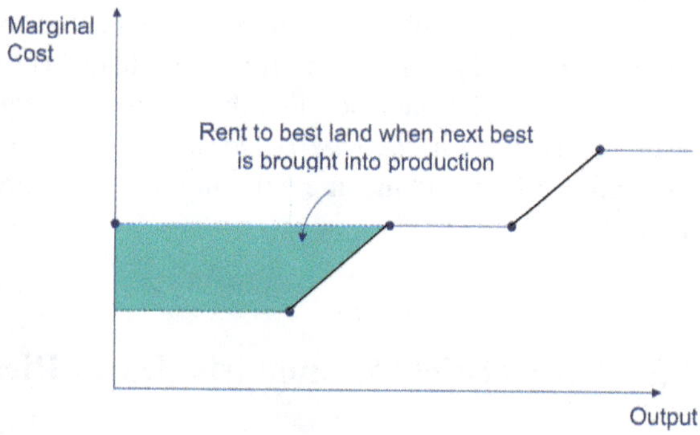


**Fig. 1.3** Ricardian rent

more and less fertile deposits, as there is more and less fertile land. The distinction among different qualities of land or resources gives rise to the concept of rent, often called Ricardian rent.

As indicated on Fig. 1.3, the owner of the good land receives a return, or rent, on the land, the difference between the cost of production from the good land and the price of the agricultural or other commodity produced, in turn given by the cost of production from the poorer, or marginal, land. This concept can of course be extended to any number of types or qualities of land or resources that can be used to produce the same commodity. The principle is that price in the market is determined by the marginal cost of production from the poorest quality, or marginal, land or resource, with owners of all infra-marginal units receiving rents. The concept of resource royalty—different but related to Ricardian rent—will also be particularly relevant when we consider the problem of optimal depletion of an exhaustible resource.

An important qualification: Diminishing returns to the only variable factor, labor, are generated as beyond some point more and more labor (the growing population) is applied to the fixed factor, land. What's left out of this story? Suppose there is another variable input, capital in all its forms. Then we can think of investing, or accumulating capital, to in effect expand the supply of the fixed factor, forestalling diminishing returns to labor. Or, we may simply think of capital as another variable factor, which can be used by labor to increase productivity. Malthus and Ricardo were in fact aware of the existence of capital, but made a very strong assumption: labor and capital are applied in "doses" of unvarying proportions—for example, one man, one plow—in effect reducing the two variable inputs to one. This is not as unrealistic an assumption as it might appear today, bearing in mind that around the end of the 18th century and early in the 19th it described fairly accurately the norm in an economy almost exclusively agricultural, and before the advent of the industrial revolution, which was to transform agriculture in very profound ways.

## 1.4 The Computation of Rent: Example

The concept of Ricardian rent can be illustrated with a numerical example. For an input of $200 of labor and farm capital (machines, fertilizer, etc.), the following yields are produced: on an acre of A (the best) land, 160 bushels of wheat; on B land, 100 bushels; on C land, 40 bushels. If all land is in production, what will be the selling price of a bushel? What is the rent to an acre of each type of land (ignoring any rent generated by increasing costs along the intensive margin)?

The marginal cost of cultivating wheat on land of type A is $200/acre/160 bushels/acre = $1.25/bushel. The marginal cost of cultivating land of type B = $200/100 = $ 2.00/bushel and land of type C is $200/40 = $5.00/bushel. If all land is in production, then the price of wheat must be at least $5.00/bushel. If the price of wheat is exactly $5.00/bushel, then farmers owning land of type A make a profit of $5.00 - $1.25 = $3.75/bushel which corresponds to a land rent of $3.75/bushel × 160 bushels/acre = $ 600/acre, and farmers with land of type B make a profit of $5.00 - $2.00 = $3.00/bushel which corresponds to a land rent of $3.00/bushel × 100 bushels/acre = $ 300/acre (Table 1.1).

**Table 1.1** Summary of computation of rent example

| Land type | Yield for $200 (bushels per acre) | Marginal cost ($ per bushel) | Profit per bushel (price $5 per bushel) | Land rent (profit per acre) |
|---|---|---|---|---|
| A | 160 | 1.25 | 3.75 | 600 |
| B | 100 | 2.00 | 3.00 | 300 |
| C | 40 | 5.00 | 0 | 0 |

## 1.5 The Classical Economists: John Stuart Mill

The last of the great classical economists, Mill made a number of innovations, presented in Principles of Political Economy in 1848 (Mill 1848), that are in some ways very close to the assumptions and features of modern models of natural (and even environmental) resource use: Mill is more Ricardian than Ricardo, in that higher costs set in even without growth in the economy and in demand for a mineral resource, as higher quality deposits are exhausted; indeed, higher costs may set in even for a single deposit as for example caused by the need to dig deeper to continue working a seam. This is a key feature of modern theories of exhaustible resource use that allow for varying extraction costs; Malthus and Ricardo did not consider exhaustion.

On the other hand, Mill is more optimistic than Malthus or Ricardo, in that he allows for sustained technical change, as opposed to the one-time improvement discussed by Malthus. Similarly, Mill allows for new discoveries (of resources), the development of new institutions, and so on, all of which mitigate scarcity—the diminishing returns or rising costs that would otherwise set in. His consideration of the continuing impact of technical change on the broader economy, as opposed to the earlier focus on a more or less static agricultural economy, is understandable as a reflection of the industrial revolution just getting going in England at the time. Mill also allows for fertility control, downplayed by Malthus and not considered by Ricardo. (Mill was a "free thinker", or nonconformist, supporting among other things women's rights, including the right to vote, and contraception. He ran for parliament and was badly beaten).

Another innovative feature of Mill's thinking is that natural resources, including land, can be valuable not only for what can be extracted from them, but for their on-site, unmodified features such as the opportunity they offer to engage in outdoor recreation or to experience natural beauty or solitude. This idea is of course very modern (in economics, if not in literature), part of the subject matter of environmental economics, as we shall discuss later on.

Somewhat surprisingly, in view of all these insights, Mill does not seem to have picked up on the idea of capital accumulation as a way of circumventing diminishing returns by adding another variable factor to the input mix, continuing to speak of a "dose" of labor and capital.

# Chapter 2
# Optimal Depletion of Exhaustible Resources

**Abstract** Exhaustible resources are non-producible and in that respect different from renewable resources, considered in the next chapter. Optimal depletion over many periods is about the tradeoff between the net benefits of extraction today and the net benefits of extraction in the future. The key insight here is that extracting a unit of the resource today carries an opportunity cost beyond the cost of the inputs used in extracting, the value that might have been obtained by extracting at some future date. The question is: What is the time pattern of extraction that maximizes the net present value of the resource in the ground? This is a constrained optimization problem, solved using elementary calculus techniques to obtain the result that the opportunity cost, the difference between price and marginal extraction cost, also known as the royalty, must grow at a rate equal to the rate of interest (here understood as the theoretically correct rate rather than an empirical rate which might be influenced by political manipulations or economic imperfections). Equivalently, the present discounted value of the royalty must be the same across all periods. If this were not so, some gain could be had by shifting a unit of extraction from a lower value period to a higher, so the initial configuration would have been neither efficient nor an equilibrium. Further results are developed in a variety of models ranging from a simple two-period model to extraction over many periods, including how to find the optimal exhaustion date for a mine, the effect of shocks and market imperfections, and finally to the case of continuous extraction. Simple algebraic and geometric analyses and many worked examples are used, along with graphical presentations. A major focus throughout is the relation between the socially efficient rate of extraction and the market-determined rate. For the continuous case, the mathematics of dynamic optimization, in particular optimal control, is introduced and applied to the resource problem, yielding additional insight into the solution. This will prove useful also in the next chapter on renewable resources such as forests, characterized by continuous growth.

Exhaustible resources are in our treatment non-producible and in that respect distinguished from renewable resources, considered in the next chapter. Optimal depletion or extraction of an exhaustible resource over many periods is about the tradeoff between benefits and costs associated with extraction today and benefits and costs



A. C. Fisher, *Lecture Notes on Resource and Environmental Economics*, The Economics of Non-Market Goods and Resources 16,
https://doi.org/10.1007/978-3-030-48958-8_2

of extraction in the future. Before proceeding with the discussion of the relevant concepts and dynamic models that address this tradeoff, we need to consider two important qualifications.

First, are resources really exhaustible, given the possibility of recycling? Clearly this does not apply to the major energy resources, the fossil fuels coal, oil and natural gas. The combustion products cannot be retrieved and used to re-constitute the original raw materials. Hardrock minerals, or metals, such as copper and aluminum, are another story. They can be and are recycled. But as Solow (1974a) has noted, this goes only so far. It is not possible, for reasons of cost, to retrieve all of the material. Some is inevitably lost in each round of recycling, so that ultimately there is a limit on how much of the resource is or will be available. Of course, given the resource, the nature of its use, and the technology for retrieval and reprocessing, it may be possible to greatly extend the apparent limit given by the stock of the resource in the ground.

The second qualification has to do with the prospects for discovery of new sources of the extractive resource. This derives from the uncertainty surrounding just how much of the mineral is in fact present in the earth's crust. Discoveries of new oil fields are not uncommon, especially in areas that have not been extensively explored. And as technology advances, it becomes possible to commercially produce, in this case oil, from unconventional and previously untapped sources, such as the oil sands in western Canada and more recently by hydraulic fracturing, or fracking, in several states including North Dakota, Texas and California. Geologists draw a distinction between reserves, the known quantity of a mineral that can be profitably extracted using current technology, and resources, the quantity that might ultimately be available as additional sources are discovered, and as the technology for discovery, extraction, and processing advances and the costs are correspondingly reduced. In either case ("reserves" or "resources"), there is in principle a limit on the stock of an exhaustible resource that is relevant to a decision on how much to extract in the current period and how much in future periods. The concept of "reserves" and its relation to what might ultimately be extracted is discussed in Sect. 5, which deals with issues of resource scarcity and sustainability.

In the remainder of this section we abstract from questions of recycling and discovery, and in our modeling exercises and numerical illustrations simply assume a known stock of the resource in question. But however we pose the problem of depletion over time, the concept of discounting, discussed below, is central, as it is in any dynamic problem or model.

## 2.1 Introduction to Discounting

Even in a world with no inflation, a dollar received a year from now should be worth less to you than a dollar received today. Say that the simple interest rate for one year is 15%. If you invest that dollar today, you would receive $\$1 \times (1 + 0.15) = \$1.15$ a year from now, so given market interest rates of 15%, you should be indifferent

**Table 2.1** Present value of $100, as a function of discount rate and year received

| T r | 1% | 5% | 10% | 15% |
|---|---|---|---|---|
| 0 | 100 | 100 | 100 | 100 |
| 1 | 99 | 95 | 91 | 87 |
| 2 | 98 | 91 | 83 | 76 |
| 3 | 97 | 86 | 75 | 66 |
| 4 | 96 | 82 | 68 | 57 |
| 5 | 95 | 78 | 62 | 50 |
| 6 | 94 | 75 | 56 | 43 |
| 7 | 93 | 71 | 51 | 38 |
| 8 | 92 | 68 | 47 | 33 |
| 9 | 91 | 64 | 42 | 28 |
| 10 | 91 | 61 | 39 | 25 |

between a dollar today and $1.15 a year from now. If you want to have $1 a year from now, you would need to invest $\$1/(1+0.15) \cong \$0.87$ today. Because capital is productive, i.e. because it yields interest, you should be indifferent between receiving $0.87 today and a dollar a year from now. Put another way, having to wait a year to receive that dollar makes it worth $0.13 cents less to you today than if you were to receive it today (and be able earn $0.13 from investing it).

The value today of payments received in the future is called *present value*. We use present values to compare cost and benefits that occur at different points in time.[1] In the above example, the present value of a dollar a year from now is $0.87.

A dollar today yields $\$1.15\ (1+0.15) = \$1.3225$ two years from now. In the second year, you receive interest on both the initial dollar investment and on the interest that accrued in the first year. A present investment of $y$ at simple interest rate $r$ yields $(1+r)y$ after one year, $(1+r)^2y$ after 2 years, and $(1+r)^Ty$ after $T$ years.

The present value of $x$ represents how much would you need to invest today in order to have $x$ available $T$ years from now. In symbols,

$$PV(x) = \left(\frac{1}{1+r}\right)^T x$$

where $T$ is the year in which you receive $x$. If $x$ is a cost that occurs in the future, you can think of the present value as the money that you would need to set aside today in order to cover that cost the day it takes place.

Interpret the bottom right entry in Table 2.1 as "at real interest rates of 15%, a cost or benefit of $100 incurred 10 years from now is worth $25 today." As shown in

[1] Our reference year does not need to be the present day—it could also be next year or 10 years from now; what is important is that all costs and benefits be valued relative to the same reference year.

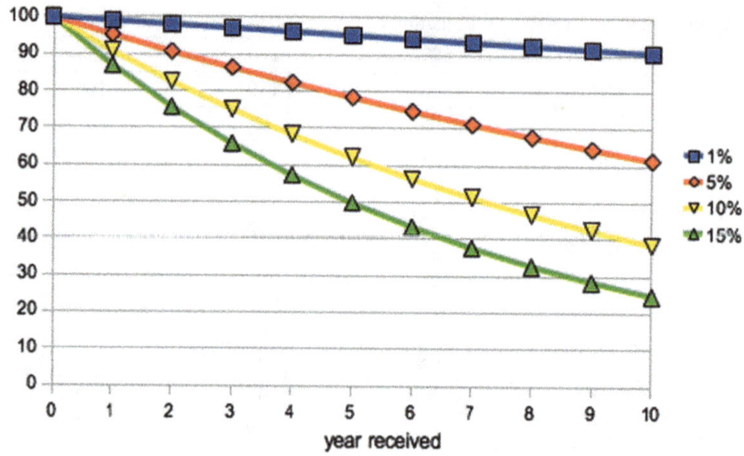


**Fig. 2.1** Present value of $100, as a function of discount rate and year received

Table 2.1 and Fig. 2.1, present values differ most from the value in the period received if interest rates are high (if $r$ is large) and if the period when the cost or benefit is incurred is far in the future (if $T$ is large).

### 2.1.1 *Estimating the Present Value of Investment Streams*

The value of a stream of costs and benefits that occur at different periods in time is the sum of the discounted individual values. The present value of an investment stream is:

$$PV = x_0 + \frac{x_1}{1+r} + \frac{x_2}{(1+r)^2} + \frac{x_3}{(1+r)^3} + \cdots$$

for all costs and benefits $x_t$ that occur at time $t$.

If the same amount $x$ is received each period until the end of time, then, recognizing a geometric series with $y = \frac{1}{1+r} < 1$, we can simplify[2]

$$PV = x + \frac{x}{1+r} + \frac{x}{(1+r)^2} + \frac{x}{(1+r)^3} + \cdots = x\left(\frac{1+r}{r}\right)$$

[2]Remember the formula for a geometric series for any $y < 1$ summed infinitely:

$$1 + y + y^2 + y^3 + \cdots = \frac{1}{1-y}$$

$$PV = x\left(1 + \left(\frac{1}{1+r}\right) + \left(\frac{1}{1+r}\right)^2 + \left(\frac{1}{1+r}\right)^3 + \cdots\right) = x\left(\frac{1}{1-\left(\frac{1}{1+r}\right)}\right) = x\left(\frac{1}{\frac{1+r-1}{1+r}}\right) = x\left(\frac{1+r}{r}\right)$$

If there is no payment in the initial period,[3] then the sum simplifies to:

$$PV = x\left(\frac{1+r}{r}\right) - x = x\left(\frac{1+r-r}{r}\right) = \frac{x}{r}$$

## 2.2 Basic Concepts and an Informal Derivation of Optimal Depletion Rates

An exhaustible resource is different from other goods or resources in that it is limited in supply and not producible. The implication of exhaustibility for efficient use or market equilibrium is that extracting and consuming a unit today carries an opportunity cost beyond the cost of the inputs used in extracting the resource: the value that might have been obtained at some future date. This opportunity cost must be considered in deciding how to allocate the resource over time. With extractible resources, the usual efficiency or equilibrium condition, price = marginal cost, can be written as:

price = marginal extraction cost + marginal opportunity cost

This equation implies that owners of a finite resource will wish to extract less today than they would if the resource were producible, as shown in Fig. 2.2, where $q^*$ is the optimal level of extraction and $q'$ is the level at which price equals marginal extraction cost.

What we have identified as the marginal opportunity cost is known by a number of names in the resource economics literature: royalty, user cost, net price, marginal profit, and scarcity rent. We'll stick with royalty, probably the most common term. Another equilibrium or efficiency condition describes the behavior of the royalty over time. Consider the decision on whether to extract a unit of the resource. The net benefit to the owner from extracting today is the royalty, as we have defined it. But that same unit would also yield a royalty if extracted next year (for now, let's assume just two periods, this year and next). So when should it be extracted to yield the largest net benefit? There is a simple and elegant answer to this question.

To see it, consider an example: For an individual producer, considering whether to extract the marginal unit of the resource this period or next, let marginal cost of extraction $MC = 2$, price $p_0 = 5$, and expected future price $p_1 = 6$. Then this year's royalty on the unit = 3, and next year's = 4. But the present value of next year's royalty $= 4/(1+r)$, where $r$ is the discount rate. Let $r = 0.1$, so $4/(1+r) = 4/1.1 = 3.64$. Extraction is optimally deferred to next year even with discounting.

[3] That is, if the investment stream is the above minus $x$:

$$\frac{x}{1+r} + \frac{x}{(1+r)^2} + \frac{x}{(1+r)^3} + \cdots$$

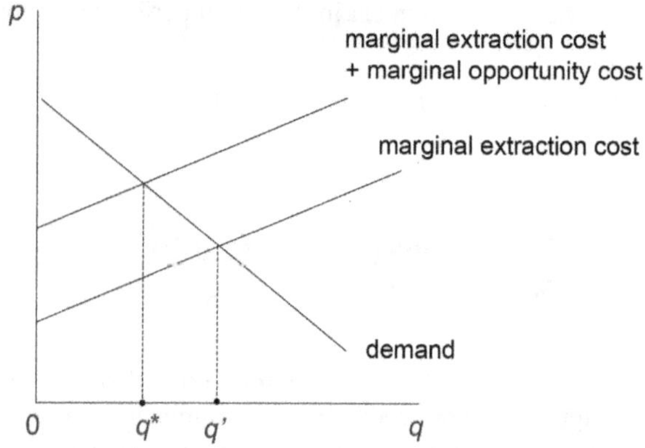


**Fig. 2.2** Optimal level of extraction $q^*$ is where price equals marginal cost, including marginal opportunity cost

Now suppose this calculation is made by many independent producers. How does the market adjust? The current price will rise, reflecting the fall in current supply. The expected future price might fall, based on what will now look like increased supplies being brought to market in the future. The cost of the marginal unit produced today may also fall if producers are confronted by a rising marginal cost curve. Any one, or a combination, of these effects will result in an increase in the current royalty relative to next year's. To keep the example simple, suppose only the current price changes, rising to 5.75. Then, the current royalty = 3.75, which is greater than next year's 3.64, and the marginal unit should be produced now.

The real question is: What structure of royalties over time is needed to eliminate the possibility of any gain in shifting extraction of the marginal unit of the resource from one period to another? In our example, what royalty is needed in the first period? Clearly, 3.64, implying $p_0 = 5.64$. More generally, the present value of the royalty must be equal across periods. In symbols,

$$p_0 - MC_0 = \frac{p_1 - MC_1}{1 + r}$$

In this form the equation is known as Hotelling's rule, in recognition of a 1931 paper by Harold Hotelling (Hotelling 1931) which first showed that to maximize social welfare the royalty (price of an extractible resource less its extraction cost) should rise over time at a rate equal to the rate of interest, or the discount rate.

It is also illuminating to look at the behavior of undiscounted royalties. In our example, the royalty rises from 3.64 to 4 or by 10%. Notice that this is just the rate of discount. Suppose we consider the resource in the ground as a capital asset. For efficiency, or equilibrium in the capital market, the return to investment in different assets must be equalized on the margin (assuming no differences in risk). For an ordinary asset, the total return on investment equals capital gain, plus dividend, less depreciation. But an exhaustible resource in the ground yields no dividend, and does not depreciate, so the return on holding this asset must come entirely as a capital gain, or rise in the value of the asset. The value of the resource in the ground is, however, just the difference between the price and the cost of extracting the resource, or the royalty. So the undiscounted royalty must rise at a rate equal to the rate of interest. In symbols,

$$p_1 - MC_1 = (p_0 - MC_0) \times (1 + r)$$

Up to this point, we have been very casual about identifying a market equilibrium with efficiency or optimality in the allocation of an exhaustible resource over time. We now need to be more precise and demonstrate why and under what conditions this is the case.

## 2.3 Formal Derivation of Optimal Depletion and the Welfare Theorem

Our strategy is to (1) set up and solve an optimization problem for a hypothetical planner seeking a socially efficient allocation of the resource, in turn accomplished by maximizing the economic surplus (consumer and producer surpluses), (2) set up and solve an optimization problem for a firm seeking to maximize profits from the resource, and (3) compare the solutions.

### *2.3.1 The Planner's Problem*

The planner's problem is shown in Fig. 2.3. The quantity to be maximized is the trapezoid, ABCD, composed of consumer surplus, AED, and producer surplus, EBCD. Note that the surplus-maximizing quantity, $q*$, is shown in Fig. 2.3 as less than the ordinary equilibrium quantity given by the intersection of demand and marginal cost or supply curves, $q'$, in accord with our informal derivation, illustrated in Fig. 2.2. We now obtain this result more formally.

The problem is in fact more complex. What the planner is seeking to maximize is the present value of surpluses, the sum over both periods, with the second period surplus appropriately discounted. In other words, it is the sum of areas ABCD in both periods, with the second period's area deflated by a discount factor, that we assume the planner is maximizing. Note, for use in the analytical formulation just below, that the area under the demand curve can be represented by the integral of the curve, from the geometric interpretation of the integral. We shall also assume that the planner determines how much of the resource to produce from each of a number of small, identical mines rather than a single large mine. This complicates the derivation but facilitates comparison with the profit-maximizing firm equilibrium.

The planner's problem can be formally stated as: maximize the present value of consumer and producer surpluses over both periods, subject to the finite stock constraint on the amount of the resource that can ever be produced. In symbols,

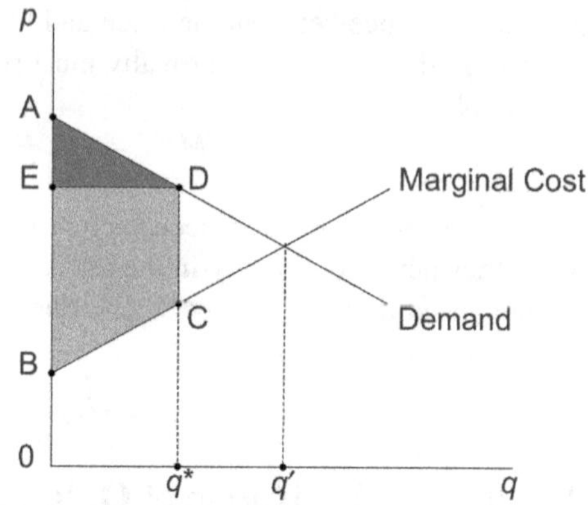


**Fig. 2.3** The hypothetical planner maximizes economic surplus at quantity extracted $q^*$

$$\max_{q_0, q_1} \int_0^{nq_0} p(q)dq - nc(q_0) + \frac{\int_0^{nq_1} p(q)dq - nc(q_1)}{1+r}$$

subject to $q_0 + q_1 \leq S$ for each of n identical deposits

We set up the Lagrangian:

$$\mathcal{L} = \int_0^{nq_0} p(q)dq - nc(q_0) + \frac{\int_0^{nq_1} p(q)dq - nc(q_1)}{1+r} + n\lambda(S - q_0 - q_1)$$

and differentiate to obtain the following first order conditions[4]:

$$(1)\ \frac{\partial \mathcal{L}}{\partial q_0} = np(nq_0) - n\frac{dc}{dq_0} - n\lambda = 0$$

$$(2)\ \frac{\partial \mathcal{L}}{\partial q_1} = \frac{np(nq_1) - n\frac{dc}{dq_1}}{1+r} - n\lambda = 0$$

$$(3)\ \frac{\partial \mathcal{L}}{\partial \lambda} = n(S - q_0 - q_1) = 0$$

[4]A note on the algebra:

$$\frac{\partial}{\partial q_0}\int_0^{nq_0} p(q)dq = \frac{\partial}{\partial q_0}\left[P(q)\,\big|_0^{nq_0}\right] \text{ where } P(q) = \int p(q)dq$$

$$= \frac{\partial}{\partial q_0}[P(nq_0) - P(0)] = \frac{\partial}{\partial q_0}P(nq_0) \text{ because } P(0) \text{ is constant}$$

$$= \frac{dP}{d(nq_0)} \cdot \frac{d(nq_0)}{dq_0} = p(nq_0)n$$

Canceling the $n$'s in Eqs. (1) and (2), the first-order conditions become

$$(1')\ p(nq_0) - \frac{dc}{dq_0} = \lambda$$

$$(2')\ \frac{p(nq_1) - \frac{dc}{dq_1}}{1+r} = \lambda$$

Combining the two equations we obtain the result:

$$p_0 - MC_0 = \frac{p_1 - MC_1}{1+r}$$

where $p_t = p(nq_t)$ and $MC_t = \frac{dc}{dq_t}$ for $t = 0, 1$.

Finally, since $n$ in Eq. (3) is positive,

$$(3')\ S - q_0 - q_1 = 0$$

### 2.3.2 The Firm's Problem

The problem to be solved by the competitive, profit-maximizing firm, the owner of one of the identical small mines, is more familiar and more straightforward. The firm's problem is to choose the quantity of the resource to be produced in each period to maximize the present value of profits, subject to the finite stock constraint it faces. In symbols,

$$\max_{q_0, q_1} p_0 q_0 - c(q_0) + \frac{p_1 q_1 - c(q_1)}{1+r} \text{ s.t. } q_0 + q_1 = S$$

$$\mathcal{L} = p_0 q_0 - c(q_0) + \frac{p_1 q_1 - c(q_1)}{1+r} + \mu(S - q_0 - q_1)$$

$$(4) \frac{\partial \mathcal{L}}{\partial q_0} = p_0 - \frac{dc}{dq_0} - \mu = 0$$

$$(5) \frac{\partial \mathcal{L}}{\partial q_1} = \frac{p_1 - \frac{dc}{dq_1}}{1+r} - \mu = 0$$

$$(6) \frac{\partial \mathcal{L}}{\partial \mu} = S - q_0 - q_1 = 0$$

Compare (4)–(6) to (1)–(3) or (1')–(3'). We have proved an extension of the basic theorem of welfare economics to the allocation of an exhaustible resource over time: A socially efficient allocation can be supported by a set of equilibrium prices, or a market equilibrium is socially efficient. This is a very powerful result, but it depends on a number of assumptions.

Before proceeding to explore the assumptions and the consequences of relaxing them, in the following two Sects. 2.4 and 2.5, we introduce a more powerful analytical approach to dynamic optimization, and use it to shed more light on the problem of optimal depletion. These sections can be skipped, and students can proceed directly to Sect. 2.6 on assumptions underlying the welfare theorem.

## 2.4 Optimal Control*

This section and the next, on optimal depletion, are based on the treatment in Conrad and Clark (1987), ch. 1, pp. 25–37 with some additional material.

Consider the problem in which an agent optimizes net economic return (utility or profit), $V(x(t), y(t), t)$, and $F(x(T))$, a scrap value function.

$$\max \int_0^T V(x(t), y(t), t)\mathrm{d}t + F(x(T))$$

$$\text{s.t. } \dot{x} = f(x(t), y(t)) \text{ and } x(0) = a \text{ (given)}$$

Here $y$ is the control variable (in the resource depletion problem, the quantity extracted in each period), and $x$ is the state variable (in the depletion problem, the stock of resource remaining).

Form the Lagrangian

$$\mathcal{L} = \int_0^T [V(\cdot) + \lambda(t)(f(\cdot) - \dot{x})]\mathrm{d}t + F(\cdot)$$

Note the analogy to the discrete time case with T periods and T corresponding constraints. Here the constraints are summed and included in the integral.

The $-\lambda_t \dot{x}$ component[5] may be integrated by parts[6] to yield $\int_0^T \dot{\lambda} x \mathrm{d}t - [\lambda_T x_T - \lambda_o x_o]$

Substituting back into $\mathcal{L}$ we get:

$$\mathcal{L} = \int_0^T [V(\cdot) + \lambda_t f(\cdot) + \dot{\lambda} x_t]\mathrm{d}t + F(\cdot) - [\lambda_T x_T - \lambda_o x_o]$$

Define the Hamiltonian

[5] I am replacing parentheses by subscripts, thus $\lambda(t)$ becomes $\lambda_t$, and so on.

[6]

$$H(x_t, y_t, \lambda_t, t) = V(\cdot) + \lambda_t f(\cdot)$$

where $V(\cdot)$ is the net flow of current benefits and $\lambda_t f(\cdot)$ is the value of the change in the state variable (which depends on future benefits).

Substitute in $\mathcal{L}$ to get

$$\mathcal{L} = \int_0^T [H(\cdot) + \dot{\lambda} x_t]\mathrm{d}t + F(\cdot) - [\lambda_T x_T - \lambda_o x_o]$$

To get the first order necessary conditions, consider a change in the control trajectory from $y_t$ to $y_t + \Delta y_t$ which causes a change in the state trajectory from $x_t$ to $x_t + \Delta x_t$. The change in the Lagrangian is:

$$\Delta\mathcal{L} = \int_0^T \left[\frac{\partial H}{\partial y_t}\Delta y_t + \frac{\partial H}{\partial x_t}\Delta x_t + \dot{\lambda}\Delta x_t\right]\mathrm{d}t + [F'(\cdot) - \lambda_T]\Delta x_T]$$

For a maximum, $\Delta\mathcal{L}$ must vanish for any $\{\Delta y_t\}$. Note that $\{\ \}$ indicates a sequence, the trajectory of $\Delta y_t$. Then the first-order conditions are:

$$\frac{\partial H}{\partial y_t} = 0 \qquad \dot{\lambda} = \frac{-\partial H}{\partial x_t} \qquad \lambda_T = F'(\cdot)$$

From the definition of $H(\cdot)$ and taking into account the initial conditions, we also have the necessary conditions:

$$\dot{x} = \frac{\partial H}{\partial \lambda_t} \quad \text{and} \quad x_o = \mathrm{a}$$

The condition on $\lambda_T$, the terminal shadow price, is known as a transversality condition (end point, after you have traversed the entire path). It says that the shadow price of the state variable at the end of the program ($\lambda_T$) is just equal to the marginal value of the state variable in the terminal or scrap value function, $F'(\cdot)$.

The other two first order conditions (on $y_t$ and $x_t$) also have very natural interpretations. I'll come back to these shortly in the framework of an optimal depletion problem.

---

$$\frac{\mathrm{d}(\lambda_t x_t)}{\mathrm{d}t} = \lambda_t \dot{x} + x_t \dot{\lambda}$$

$$\int_0^T \frac{\mathrm{d}(\lambda_t x_t)}{\mathrm{d}t}\mathrm{d}t = \int_0^T \lambda_t \dot{x}\mathrm{d}t + \int_0^T \dot{\lambda} x_t \mathrm{d}t$$

$$-\int_0^T \lambda_t \dot{x}\mathrm{d}t = \int_0^T \dot{\lambda} x_t \mathrm{d}t - \int_0^T \frac{\mathrm{d}(\lambda_t x_t)}{\mathrm{d}t}\mathrm{d}t = \int_0^T \dot{\lambda} x_t \mathrm{d}t - \lambda_t x_t \Big|_0^T$$

$$= \int_0^T \dot{\lambda} x_t \mathrm{d}t - [\lambda_T x_T - \lambda_o x_o].$$

First, let's develop another transversality condition, which will be relevant to the depletion problem. Ordinarily, in such a problem, the terminal time is free, i.e., $T$ is not known—in fact, finding $T$ is a big part of the problem (when do we (optimally) run out of the resource?)

To find the optimal $T$, set $\frac{\partial \mathcal{L}}{\partial T} = 0$

$$\frac{\partial \mathcal{L}}{\partial T} = H_T + \dot{\lambda}_T x_T + F'(\cdot)\dot{x}_T - \dot{\lambda}_T x_T - \lambda_T \dot{x}_T = 0$$

$$\Rightarrow H_T = 0 \quad \text{since} \quad F'(\cdot) = \lambda_T$$

This is the other relevant transversality condition, which I'll interpret in the depletion problem.

The maximum principle is the set of first order conditions plus the transversality conditions:

$$\frac{\partial H}{\partial y_t} = 0 \qquad \dot{\lambda} = -\frac{\partial H}{\partial x_t} \qquad \lambda_T = F'(\cdot) \qquad H_T = 0$$

Note: If control constraints such as $y_t \in Y$ exist, there may not be an interior solution for $y_t$. In this case, the condition $\frac{\partial H}{\partial y_t} = 0$ takes the more general form: $y_t$ maximizes $H(x_t, y_t, \lambda_t, t)$ over $y_t \in Y$ for $0 \leq t \leq T$ (Boltyanskiy et al. 1962).

Example: Say $H$ is linear in the control variable. Then $H = pq - cq - \lambda q$, $\frac{\partial H}{\partial q} = p - c - \lambda$, which implies that $q = \bar{q}$ if $p > c + \lambda$ where $\bar{q}$ = maximum production, and $q = 0$ if $p < c + \lambda$.

### 2.4.1 Discounting

Discounting could be considered implicit in our specification of $V(x_t, y_t, t)$ since $V$ is a function of $t$—though $F(\,)$ was not written as a function of $T$, which it should be if discounted. Let's now introduce discounting explicitly:

$$\max \int_0^T V(x_t, y_t)e^{-rt}dt + F(x_T)e^{-rT}$$

$$s.t.\ \dot{x} = f(x_t, y_t) \text{ and } x_0 = \text{a}$$

We write the Hamiltonian as

$$H = V(\cdot)e^{-rt} + \lambda_t f(\cdot)$$

(Note: $\lambda_t$ is a present value of future changes)

Define the current value Hamiltonian:

$$\tilde{H} = He^{rt} = V + \mu_t f(\cdot) \quad \text{where } \mu_t = \lambda_t e^{rt}$$

We can now rewrite the maximum principle in terms of $\tilde{H}$:

$$\frac{\partial H}{\partial y_t} = \frac{\partial V}{\partial y_t} e^{-rt} + \lambda_t \frac{\partial f}{\partial y_t} = 0$$

$$\dot{\lambda} = -\frac{\partial H}{\partial x_t} = -\frac{\partial V}{\partial x_t} e^{-rt} - \lambda_t \frac{\partial f}{\partial x_t}$$

From the definition of $\mu_t$:

$$\lambda_t = \mu_t e^{-rt} \text{ and } \dot{\lambda}_t = -r\mu_t e^{-rt} + \dot{\mu} e^{-rt}$$

substituting these expressions in the above equations,

$$\frac{\partial V}{\partial y_t} e^{-rt} + \mu_t e^{-rt} \frac{\partial f}{\partial y_t} = 0 \rightarrow \frac{\partial V}{\partial y_t} + \mu_t \frac{\partial f}{\partial y_t} = 0$$

$$-r\mu_t e^{-rt} + \dot{\mu} e^{-rt} = -\frac{\partial V}{\partial x_t} e^{-rt} - \mu_t e^{-rt} \frac{\partial f}{\partial x_t} \rightarrow -r\mu_t + \dot{\mu} = -\frac{\partial V}{\partial x_t} - \mu_t \frac{\partial f}{\partial x_t}$$

$$\rightarrow \dot{\mu} = r\mu_t - \frac{\partial V}{\partial x_t} - \mu_t \frac{\partial f}{\partial x_t}$$

In terms of the current value Hamiltonian,

$$\text{maximum principle} \begin{cases} \dfrac{\partial \tilde{H}}{\partial y_t} = 0 \\ \dot{\mu} = r\mu_t - \dfrac{\partial \tilde{H}}{\partial x_t} \\ \text{and of course } \dot{x} = \dfrac{\partial \tilde{H}}{\partial \mu_t}, \ x_o = \text{a, and} \begin{Bmatrix} \mu_T = F'(\cdot) \\ \tilde{H}_T = 0 \end{Bmatrix} \text{transversality conditions} \end{cases}$$

## 2.5 Optimal Depletion Using Optimal Control*

### 2.5.1 *Efficient Allocation*

The planner's problem can be written as

$$\max \int_0^T [U(q_t) - c(q_t, x_t)] e^{-rt} dt$$
$$\text{s.t. } \dot{x} = -q_t, \ x(0) \text{ given}$$

where $q_t$ is the control variable, the quantity extracted in period $t$.

The current-value Hamiltonian is

$$\tilde{H} = U(q_t) - c(q_t, x_t) - \mu_t q_t$$

$$\begin{aligned} \text{where } U(q_t) - c(q_t, x_t) &= \text{net flow benefit} \\ -\mu_t q_t &= \text{value of investment, in this case} \\ &\quad \text{the shadow price of a unit in} \\ &\quad \text{the stock, } \mu, \text{ times the} \\ &\quad \text{change in the stock, } -q \end{aligned}$$

So $H$ summarizes the net impact of a choice of control at any instant, made up of two components, the current (rate of) flow plus the impact on future flow.

First-Order Conditions are

$$\frac{\partial \tilde{H}}{\partial q_t} = U'(q_t) - \frac{\partial c}{\partial q}(q_t, x_t) - \mu_t = 0 \qquad (1)$$

$$\text{If } U(q) = \int_0^q p(z)\mathrm{d}z = \text{ area under the demand curve from 0 to q)}$$

then (1) becomes:

$$\frac{\partial \tilde{H}}{\partial q} = p(q_t) - \frac{\partial c}{\partial q}(q_t, x_t) - \mu_t = 0 \qquad (1)'$$

$$\left(\text{since } \frac{\mathrm{d}}{\mathrm{d}q}\int_0^q p(z)\mathrm{d}z = \frac{\mathrm{d}}{\mathrm{d}q}P(z)\Big|_0^q = \frac{\mathrm{d}}{\mathrm{d}q}[P(q) - P(0)] = \right.$$

$$\left. = \frac{\mathrm{d}}{\mathrm{d}q}P(q), \text{ since } P(0) \text{ is constant, } = p(q)\right)$$

where $x_t$ is the state variable, representing the remaining stock of resource.

$$\dot{\mu} = r\mu + \overbrace{\frac{\partial c}{\partial x}}^{<0} \qquad (2)$$

But note that if $c = c(q)$ get :

$$\dot{\mu} = r\mu - \frac{\partial \tilde{H}}{\partial x} = r\mu \text{ since } \frac{\partial \tilde{H}}{\partial x} = 0 \qquad (2)'$$

Alternatively, if $U(q, x)$, that is if utility is derived from the resource in the ground, as described in the preceding chapter, then we may also have $\dot{\mu} < r\mu$ since:

$$\dot{\mu} = r\mu - \overbrace{\frac{\partial U}{\partial x}}^{>0} \qquad (2)''$$

So, if $x$ enters the objective function, either through utility or extraction cost, then we have $\dot{\mu} < r\mu$.

We can interpret the first-order conditions:

1. (1) Consider the marginal unit extracted at t. The net flow benefit, $U'(q) - \frac{\partial c}{\partial q}(q, x)$, is just equal to the loss in future value, $-\mu$, from having one less unit in the stock
2. $(1)'$ Price is not equal to marginal cost. Instead, price equals marginal cost $\frac{\partial c}{\partial q}(q, x)+$ royalty $\mu$.
3. (2) The value of the resource in ground grows at less than rate $r$, since it also yields a dividend in the form of reduced future extraction cost, $\frac{\partial c}{\partial x}$, or utility, $\frac{\partial U}{\partial x}$, as in $(2)''$, or both.
4. $(2)'$ The value of the resource in the ground, the royalty, must grow at rate $r$, to yield the same return as other assets

Note: We could have $\dot{\mu} < 0$ and indeed $\mu$ could approach 0 if cost keeps increasing as the stock is depleted, to the point where nothing is left but "average rock" with no scarcity rent.

### 2.5.2 *Competitive Equilibrium*

The firm's problem is

$$\max \int_0^T [p_t q_t - c(q_t, x_t)]e^{-rt} \mathrm{d}t$$
$$\text{s.t.} \quad \dot{x} = -q_t, \; x(0) \; \text{given}$$

Note: Casual treatment of units; could have $q_i$ for each firm, with $c_i(q_i)$ and $\sum q_i = q$ – but it doesn't affect results

$$\tilde{H}_t = p_t q_t - c(q_t, x_t) - \mu_t q_t$$

Note: Casual treatment of symbol for costate variable $\mu$; could have $\mu_i$, but from the first necessary condition $\mu_i = \mu = p - c'(q)$, at least in equilibrium.

First-Order conditions are

$$\left.\begin{aligned} &(3)\ \frac{\partial \tilde{H}}{\partial q_t} = p_t - \frac{\partial c}{\partial q}(q_t, x_t) - \mu_t = 0 \\ &(4)\ \dot{\mu} = r\mu - \frac{\partial \tilde{H}}{\partial x} = r\mu \ \text{ since } \ \frac{\partial \tilde{H}}{\partial x} = 0 \end{aligned}\right\} \ \text{same as } (1)' \text{ and } (2)$$

An alternative way of writing (2) or (4) is "Hotelling's Rule" , defined in our analysis of the two-period problem:

$$\frac{\mathrm{d}(p - \frac{\partial c}{\partial q}(q_t, x_t))}{\mathrm{d}t} = r(p - \frac{\partial c}{\partial q}(q_t, x_t)); \ \text{if } \ \frac{\partial c}{\partial q}(q_t, x_t) = 0, \frac{\mathrm{d}p}{\mathrm{d}t} = rp \ \text{ or } \ \frac{\dot{p}}{p} = r$$

This is in fact the way the result is typically stated, though we've seen, a falling shadow price is possible if the extraction cost depends on the remaining stock, as in $c(q, x)$

Does this prove a competitive equilibrium is socially efficient? Yes, but only on the basis of a number of implicit assumptions involving market imperfections, externalities, or other market failures. These are generally well known, but apply with special force in the case of exhaustible resources, as we'll see.

Before considering the "standard" assumptions and what happens when they fail, let me note another, involving a nonconvexity, as analyzed by Fisher and Karp (1993). This is worth doing because it will also shed light on the maximum principle, and especially the role of transversality in a depletion problem.

### 2.5.3 Equilibrium, Efficiency, and Non-convexity

Suppose we have the ordinary U-shaped average cost curve, AC in Fig. 2.4, with the corresponding marginal cost curve MC. Note that this presumably involves a nonconvexity (of the total cost curve, TC, as in Fig. 2.5).

Now consider the transversality condition for a free terminal time problem (a key point of the problem is to find the time to exhaustion). For the planner (efficient allocation) this is

$$\tilde{H}_T = U(q_T) - c(q_T) - \mu_T q_T = 0$$

$$\text{so} \quad \frac{U(q_T)}{q_T} - \frac{c(q_T)}{q_T} - \bar{}_T = 0$$

But from the maximum principle, the first order condition, Eq. (1), must hold at all points along the optimal trajectory, including at $t = T$:

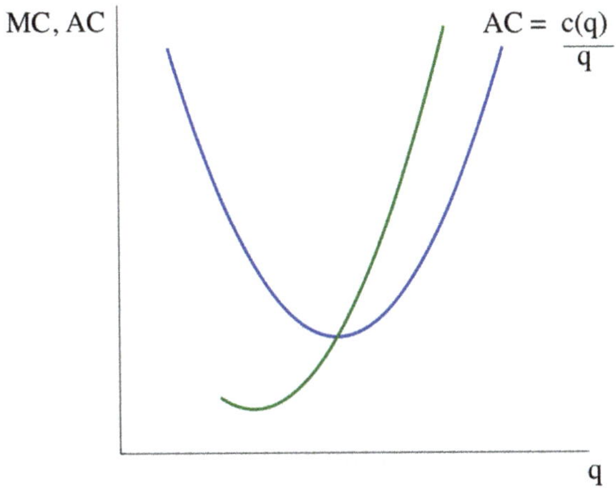


**Fig. 2.4** Non-convexity: ordinary U-shaped average cost curve

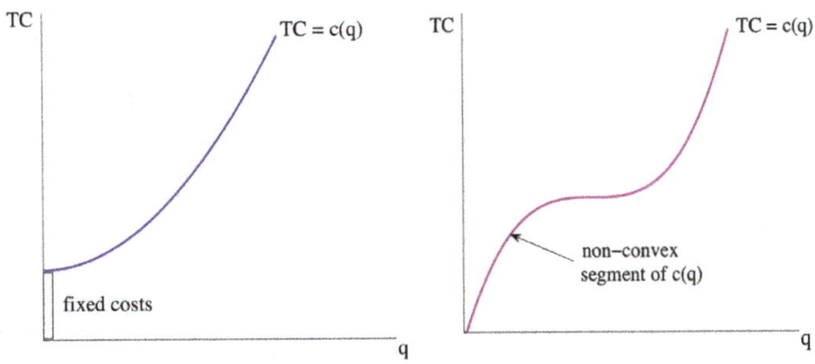


**Fig. 2.5** Associated non-convex total cost curves

$$\frac{\partial \tilde{H}}{\partial q_T} = U'(q_T) - c'(q_T) - \mu_T = 0$$
$$\Rightarrow \frac{U(q_T)}{q_T} - \frac{c(q_T)}{q_T} = U'(q_T) - c'(q_T) = p_t - c'(q_T) \text{ if } U = \int_0^{q_T} p(z)\mathrm{d}z$$

Alternatively,
$$\underbrace{\frac{U(q_T)}{q_T}}_{\text{average utility}} - \underbrace{p_T}_{\text{marginal utility}} = \frac{c(q_T)}{q_T} - c'(q_T)$$

For a concave utility function, such as $U = \int_0^q p(z)\mathrm{d}z$, average utility > marginal utility, as in Fig. 2.6.

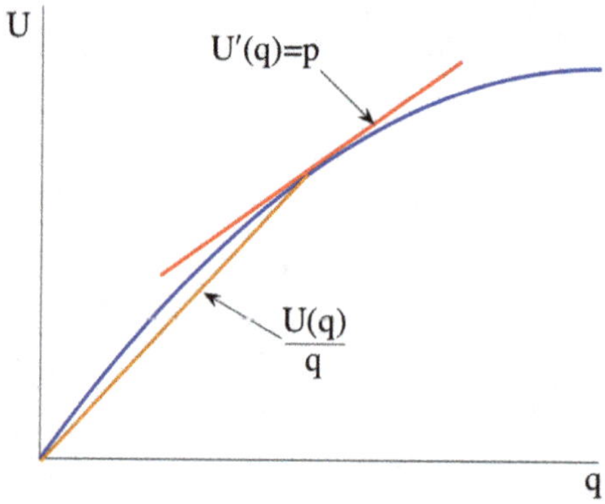


**Fig. 2.6** Concave utility function: average utility > marginal utility

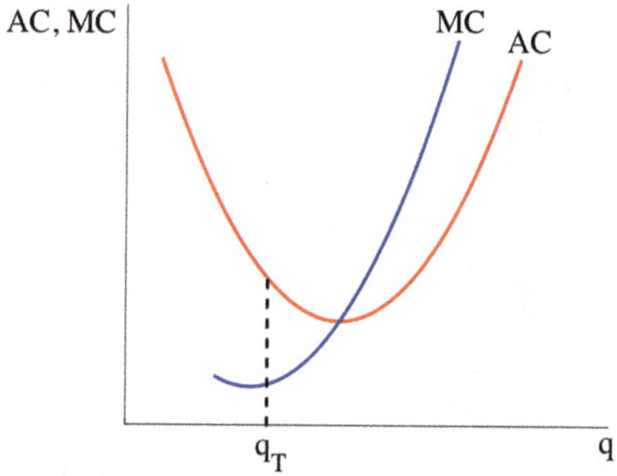


**Fig. 2.7** Production ceases at $q_T$, where AC > MC

Since the left hand side, $\dfrac{U(q_T)}{q_T} - p_T$, is greater than zero and the right hand side, $\dfrac{c(q_T)}{q_T} - c'(q_T)$, is also greater than zero, production ceases at a point where AC > MC, i.e., to the left of minimum AC, as in Fig. 2.7.

This result is driven by the transversality condition $\tilde{H}_T = 0$ which states that production ceases when the net benefit, taking into account both the current flow and the impact on future flows, falls to zero. We can gain an appreciation of this result by considering a couple of steps along the way.

Figure 2.8 illustrates that the current-value Hamiltonian is positive at a point like q' along the optimal trajectory, where q' is to the right of minimum AC. On Fig. 2.9, at a point further along the optimal trajectory, in particular q", at minimum AC, the Hamiltonian is still positive. In standard treatments, as in (Conrad and Clark 1987), production is shown to cease as the value of the Hamiltonian falls to zero at q". Clearly the standard result is inconsistent with ours, and as we shall show depends on the absence of a non-convexity. Finally, Fig. 2.10 shows that in our analysis the

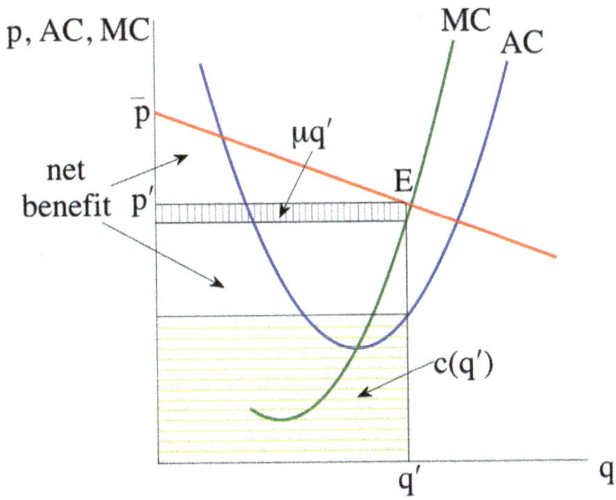


**Fig. 2.8** At a point like $q'$, where still (optimally) producing, $\tilde{H} = U(q') - c(q') - \mu q'$. Note: $U(q') = 0$, $\bar{p}Eq'$ so $U(q') > c(q') + \mu q'$ and $\tilde{H} > 0$

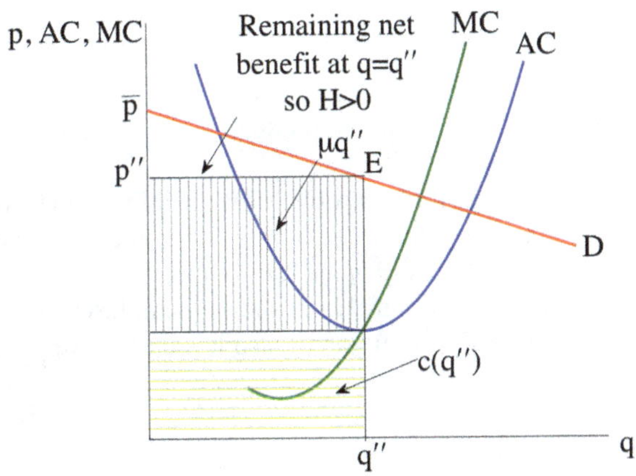


**Fig. 2.9** At $q''$, where $q'' = q_{\text{minAC}}$, the Hamiltonian is still positive

value of the Hamiltonian falls to zero only at a point like $q^*$, to the left of minimum AC.

Before proceeding to work through the implications of the non-convexity we first need to develop the conditions of the competitive equilibrium.

$$\tilde{H}_T = p_T q_T - c(q_T) - \mu_T q_T = 0$$
$$p_T - \frac{c(q_T)}{q_T} - \mu_T = 0$$

But Eq. (3) also holds at $t = T$:

$$\frac{\partial \tilde{H}}{\partial q_T} = p_T - c'(q_T) - \mu_T = 0$$

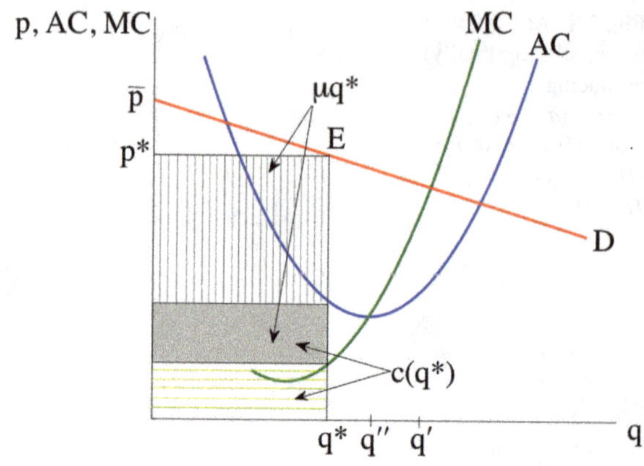


**Fig. 2.10** Shaded area is part of $\mu q^*$ and $c(q^*)$, so it gets counted twice; balanced against benefit triangle at top of figure

$$\Rightarrow \frac{c(q_T)}{q_T} = c'(q_T)$$

so production ceases where AC = MC, the point of minimum AC, different from the efficient allocation, where production ceases at a point to the left of minimum AC, as in Fig. 2.10.

But there is a further problem: the equilibrium in fact does not exist. Why? Because there is a price jump at $T$, from $p_T$ to some point further "up" the demand curve, perhaps to $\bar{p}$, where $\bar{p}$ is the "choke price".

In Fig. 2.11 the right panel shows the industry demand curve and aggregate cost curves; each firm (left panel) also has cost curves of this shape. We assume firms

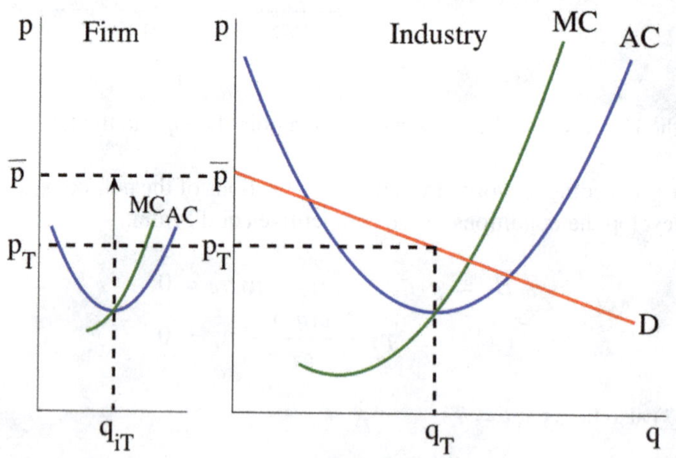


**Fig. 2.11** Firm and industry demand and aggregate cost curves

know the industry demand, as well as remaining stock of the resource, so they know there will be a jump, and when it will occur.

The point is that if firms are price takers and anticipate the jump (and we are implicitly assuming perfect information), they will want to hold on to some of their stock, i.e., hold off producing at $T$, say to $T + \Delta T$, to take advantage of the higher price. This result was first obtained by Eswaran et al. (1983) and subsequently by Fisher and Karp (1993), who go on to consider the implications of non-convexity for social efficiency and ways to restore the "standard" efficiency and equilibrium results.

There are a couple of ways to restore efficiency, equilibrium, or both. One involves the existence of a substitute or "backstop" for the resource available at a sufficiently low price, as discussed in Fisher and Karp and very briefly below.

The other is of course if the nonconvexity is eliminated, i.e., if the AC curve is not U-Shaped.

Efficient Allocation

$$\begin{aligned}
&\text{transversality} && \frac{U(q_T)}{q_T} - \frac{c(q_T)}{q_T} - \mu_T = 0 \\
&\text{max principle} && U'(q_T) - c'(q_T) - \mu_T = 0 \\
& && \Rightarrow \frac{U(q_T)}{q_T} - p_T = \frac{c(q_T)}{q_T} - c'(q_T)
\end{aligned}$$

Since $\frac{U(q_T)}{q_T} > p_T$ for all $q_T > 0$ and since $\frac{c(q_T)}{q_T} < c'(q_T)$ for all $q_T > 0$, we must have $q_T = 0$ (strictly, producing the last unit, or fraction of a unit, at $T$) where:

$$\frac{U(q_T)}{q_T} - p_T = \frac{c(q_T)}{q_T} - c'(q_T) = 0$$

This is illustrated in Fig. 2.12. In terms of $\tilde{H}_T$: as $q$ decreases from $q'''$ toward $q = 0$, the two net benefit triangles get smaller, and ultimately vanish at $q = 0 \Rightarrow \tilde{H}_T = 0$.

Competitive Equilibrium

$$\begin{aligned}
&\text{transversality} && p_T - \frac{c(q_T)}{q_T} - \mu_T = 0 \\
&\text{max principle} && p_T - c'(q_T) - \mu_T = 0 \\
& && \Rightarrow \frac{c(q_T)}{q_T} = c'(q_T) \\
& && \text{which occurs only at } q = 0 \text{ (or for last unit)}
\end{aligned}$$

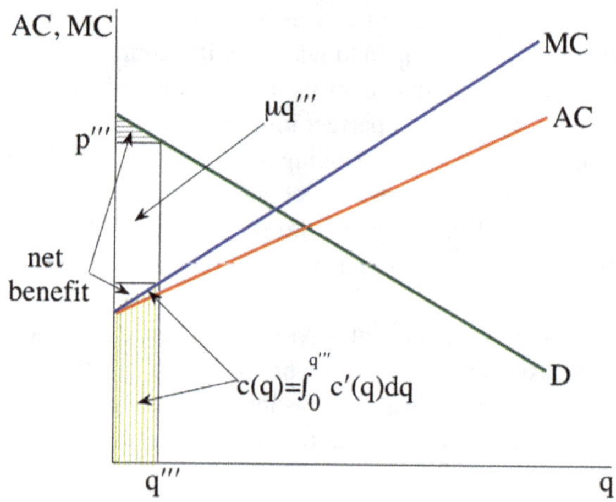


**Fig. 2.12** Efficient allocation: as $q$ decreases from $q'''$ toward $q = 0$, the two net benefit triangles get smaller, and ultimately vanish at $q = 0$

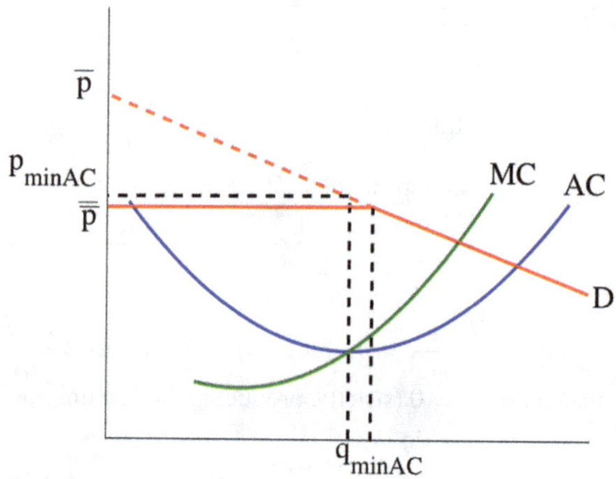


**Fig. 2.13** Substitute available

So in both cases, production ceases at point of min AC, which is in fact at $q_T = 0$, an equilibrium exists, and is consistent with efficient allocation. There is no room for price to jump at $T$, because price has reached $\bar{p}$, the choke price. Why isn't the nonconvexity a problem ordinarily? Because it is assumed output is large relative to output at min AC, i.e., firms operating where MC > AC and AC rising. It only becomes a problem with exhaustible resource extraction because output is falling over time, as price is rising, and may become small relative to the min AC output.

Now let's look at the effect of a substitute or backstop for the resource, available at a price below $\bar{p}$, and indeed below the price for the minimum AC quantity, as shown in Fig. 2.13.

Say the price of a substitute $= \bar{\bar{p}}$. In this case $p$ at $T$ remains at $\bar{\bar{p}}$. Since there is no jump, there is no benefit to firms from holding over some production to $T + \Delta T$. If the price of the substitute, $\bar{\bar{p}}$, were higher than $p_{\text{minAC}}$, of course, we would be back in the situation of an implied jump and no equilibrium.

As a final note, still another possible resolution is that if firms are uncertain about others' stocks, this implies that they can't predict the jump, or more precisely, when the jump will occur. The fact that firms can't predict the jump restores existence, but probably not efficiency.

## 2.6 Assumptions Underlying the Welfare Theorem

### 2.6.1 *Perfect Competition*

We have implicitly assumed perfect competition, since the firm is a price taker. Yet this is probably not a good assumption for many resource firms. Why? Think about the spatial distribution of deposits. For an ordinary good or service the existence of extra-normal returns is a signal to others to enter the market. But the limited number of deposits that can be profitably extracted at any given time acts as a natural barrier to entry.

Suppose the resource firm is a monopoly. The problem is exactly the same as for the competitive firm but, now, the price is affected by the quantity the firm produces in any period.[7] In symbols,

$$\max_{q_0, q_1} p_0 q_0 - c(q_0) + \frac{p_1 q_1 - c(q_1)}{1+r} \text{ s.t. } q_0 + q_1 = S$$

$$L = p_0 q_0 - c(q_0) + \frac{p_1 q_1 - c(q_1)}{1+r} + \mu(S - q_0 - q_1)$$

$$(4) \frac{\partial L}{\partial q_0} = p_0 + q_0 \frac{dp_0}{dq_0} - \frac{dc}{dq_0} - \mu = 0$$

$$(5) \frac{\partial L}{\partial q_1} = \frac{p_1 + q_1 \frac{dp_1}{dq_1} - \frac{dc}{dq_1}}{1+r} - \mu = 0$$

so

$$MR_0 - MC_0 = \frac{MR_1 - MC_1}{1+r}$$

In the monopoly equilibrium, price is replaced by marginal revenue. Thus, the relationship of the rate of depletion to that under perfect competition depends on the nature of demand. For typical demand curves, such as linear, or log linear (constant elasticity), it can be shown that, under monopoly, the resource will be depleted more slowly or, in other words, as someone, perhaps Hotelling, remarked, the monopolist

[7] See Hanley et al. (1997), ch. 9 for discussion of imperfect competition of various kinds: monopoly, duopoly, cartel.

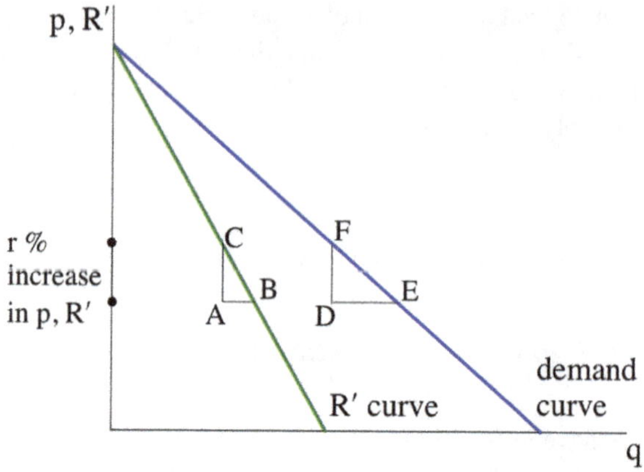


**Fig. 2.14** Competitive versus monopoly depletion

is the conservationist's friend. In the next (advanced) section, we consider the case of constant elasticity.

Results in the linear case, with elasticity decreasing over quantity, can be illustrated graphically. In Fig. 2.14, for DF = AC, i.e., equal ($r\%$) increase in $p$ and in $R'$, DE > AB so the monopolist decreases q by a smaller amount for an $r\%$ increase in $R'$ than competitive industry does for an $r\%$ increase in $p$. This implies that the extraction path will be flatter for the monopolist, with less extracted in the first or early periods and more in the second or later periods. Note: Lewis (1976) proves this result for any decreasing elasticity demand curve.

The analysis and the result, even in the relatively simple case of monopoly, are more complicated than for an ordinary good. In that case, replacement of price by marginal revenue dictates a lower output and higher price under monopoly. For an exhaustible resource, by assumption, the entire stock of the resource will ultimately be produced. If less is produced by the monopoly early on, more will be produced later.

Many resource industries are in fact characterized by an intermediate market structure: not perfectly competitive but not a monopoly. These are resource cartels (several producers band together to act like a monopoly, often with a "competitive fringe") or oligopolies (several large producers who interact with each other). These market structures are still more complicated, and more difficult to analyze, though there is a large literature given their empirical importance. No doubt the best-known example of a resource cartel with a competitive fringe is OPEC, the grouping of oil producers. For an early analysis that combines depletion theory with empirical information on demand, supply, and costs in the world oil market, see Hnyilicza and Pindyck (1976).

### 2.6.2 *Advanced**

Again, suppose the resource is exploited by a monopoly. The control problem is

$$\max \int_0^T [p_t q_t - c(q_t)]e^{-rt}\mathrm{d}t$$
$$\text{s.t. } \dot{x} = -q_t$$

$$\tilde{H} = p_t q_t - c(q_t) - \mu_t q_t$$

$$(1) \quad \frac{\partial \tilde{H}}{\partial q} = p_t + q_t \frac{\mathrm{d}p}{\mathrm{d}q} - c'(q) - \mu_t = 0$$

Note : $p_t + q_t \dfrac{\mathrm{d}p}{\mathrm{d}q}$ = marginal revenue $R'(q)$, where total revenue $R = pq$

$$(2) \quad \dot{\mu} = r\mu_t \text{ where from Eq. (1), } \mu_t = \left(p_t + q_t \frac{\mathrm{d}p}{\mathrm{d}q}\right) - c'(q)$$

Special Case: elasticity $\epsilon$ = constant and $c'(q) = 0$ (Stiglitz 1976)

$$R' = p\left(1 + \frac{1}{\epsilon}\right) \text{ where } p\left(1 + \frac{1}{\epsilon}\right) = p\left(1 + \frac{\mathrm{d}p}{\mathrm{d}q}\frac{q}{p}\right)$$
$$= p + q\frac{\mathrm{d}p}{\mathrm{d}q} = R'$$

$$= kp \text{ where } k \text{ is constant } \left(= 1 + \frac{1}{\epsilon}\right)$$

since $\mu = R'$, from Eq. (1) and $c'(q) = 0$, we can rewrite Eq. (2) as:

$$\frac{\mathrm{d}(R')}{\mathrm{d}t} = rR'$$
$$\frac{\mathrm{d}(kp)}{\mathrm{d}t} = rkp$$
$$k\frac{\mathrm{d}p}{\mathrm{d}t} = krp \text{ and } \frac{\mathrm{d}p}{\mathrm{d}t} = rp, \text{ the same condition as for the competitive industry}$$

But if $c'(q) \neq 0$, we have

$$\frac{\mathrm{d}(R' - c')}{\mathrm{d}t} = r(R' - c')$$

$$\text{If} \quad \frac{\mathrm{d}c'}{\mathrm{d}t} = 0,$$

$$\frac{\mathrm{d}R'}{\mathrm{d}t} = r(R' - c')$$

$$\frac{\mathrm{d}(kp)}{\mathrm{d}t} = r(kp - c')$$

$$k\frac{\mathrm{d}p}{\mathrm{d}t} = r(kp - c')$$

$$\frac{\mathrm{d}p}{\mathrm{d}t} = r\left(p - \frac{c'}{k}\right)$$

and since $k < 1$, $\frac{c'}{k} > c'$, and $\left(p - \frac{c'}{k}\right) < (p - c')$.

Since the competitive equilibrium is: $\frac{\mathrm{d}(p - c')}{\mathrm{d}t} = r(p - c')$, which for $\frac{\mathrm{d}c'}{\mathrm{d}t} = 0$ reduces to

$$\frac{\mathrm{d}p}{\mathrm{d}t} = r(p - c'),$$

price is growing more slowly for the monopolist, and the extraction path is also flatter, as shown in Fig. 2.15.

Note: If $\frac{\mathrm{d}c'}{\mathrm{d}t} \neq 0$, $\frac{\mathrm{d}p}{\mathrm{d}t} = r\left(p - \frac{c'}{k}\right) + \frac{\frac{\mathrm{d}'}{c}}{\mathrm{d}t}k$, then if $\frac{\mathrm{d}c'}{\mathrm{d}t} < 0$, $\left|\frac{\frac{\mathrm{d}'}{c}}{\mathrm{d}t}k\right| > \left|\frac{\mathrm{d}c'}{\mathrm{d}t}\right|$, which implies that monopoly $\frac{\mathrm{d}p}{\mathrm{d}t}$ is further reduced relative to competitive $\frac{\mathrm{d}p}{\mathrm{d}t}$.

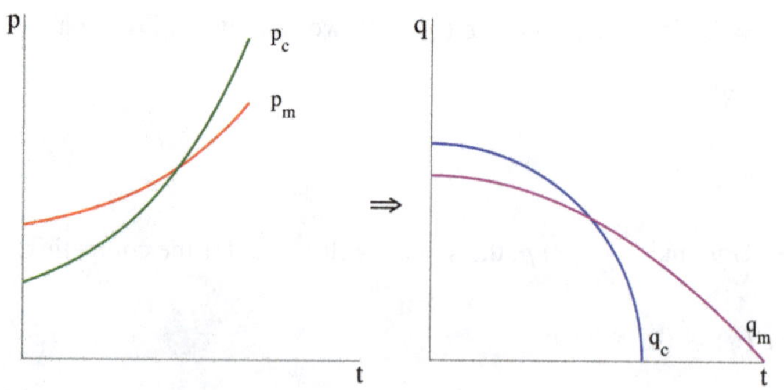


**Fig. 2.15** Competitive and monopoly price and extraction paths

### 2.6.3 No Externalities

We have implicitly assumed no externalities since marginal cost is specified in the same way for both planner and firm. This is probably also not a good assumption for the resource sector since many environmental impacts are associated with the extraction, transportation, and processing of minerals, especially energy minerals. For example, most air pollutants regulated by EPA are related to fossil fuel use, and EPA has recently been told by the U.S. Supreme Court to regulate the main greenhouse gas, carbon dioxide, produced by burning fossil fuels.

Suppose there is an external cost associated with production of the resource so that marginal social cost is above marginal private cost. What is the effect on the efficient or optimal rate of depletion?

Let the external cost be represented by $E$. Then,

$$p_0 - (MC_0 + E) = \frac{p_1 - MC_1 - E}{1 + r}$$

$$p_0 - MC_0 - E = \frac{p_1 - MC_1}{1 + r} - \frac{E}{1 + r}$$

For $r > 0$, $\frac{E}{1+r} < E$, which implies that the present value of the net benefit of extracting a unit of the resource in the second period is reduced by less than the net benefit of extracting in the first. Therefore some extraction is optimally shifted from the first period to the second. Depletion is shifted, on the margin, to the future since this reduces the present value of costs. Alternatively, we can say that market-determined depletion, without internalization of the externality, will be too rapid. In the next section, we'll illustrate this concept with a numerical example.

### 2.6.4 Advanced*

Suppose all costs are external. Then comparing the socially efficient extraction path with market-determined depletion, where $E$ represents the external costs, we have

Social:

$$\frac{\mathrm{d}(p - c')}{\mathrm{d}t} = r(p - c')$$

again, suppose $\frac{\mathrm{d}c'}{\mathrm{d}t} = 0$

then $\frac{\mathrm{d}p}{\mathrm{d}t} = r(p - c')$

Private:

$$\frac{\mathrm{d}p}{\mathrm{d}t} = rp$$

so the price path is steeper, and the resource is depleted too quickly

If $\frac{\mathrm{d}c'}{\mathrm{d}t} < 0$, $\frac{\mathrm{d}p}{\mathrm{d}t} = r(p - c') + \frac{\mathrm{d}c'}{\mathrm{d}t} < r(p - c')$

Note: This analysis assumes the only control or choice variable is the rate of extraction.

### 2.6.5 Private and Social Discount Rates Are the Same

We have implicitly assumed discount rates are the same for the planner and the firm since both are represented by the same symbol, $r$. Once again, this is probably not a good assumption, though the issue is controversial, and there is a large literature. Many students of the problem argue that the market rate is higher, because private economic agents are more risk averse than society is, or ought to be, in evaluating returns to an investment (such as, in this case, leaving resources in the ground). Another way of putting this is that part of the return to private investment is for risk-bearing, and risk in the public sector is lower. There are two sources of this view. One, associated with Samuelson (informal discussion in May Papers and Proceedings issue of the AER, 1964), is risk pooling. The government makes a very large number of investments, so if it undertakes only those for which the expected value is positive, the law of large numbers will suffice to keep the return on the average investment positive. The other is risk sharing, in that a very large number of voters/taxpayers share the risk. Arrow and Lind (1970) show that the aggregate risk premium goes to zero in the limit even as individuals display risk aversion. A subsequent study by James (1975) considers the conditions under which one view or the other is correct. A counter-argument here is that capital markets can provide both risk-pooling (think mutual funds) and risk-sharing (many shareholders in a large company). It seems safe to say that, despite decades of debate, this is not a settled issue.

Another justification for a social rate below the market, formalized by Marglin (1963a, b) is that the welfare of future generations is a public good externality. Each individual acting alone will save and invest too little, so that collective action is needed to increase the supply of funds for investment, resulting in a lower discount rate. Note, however, that the prescription is to drive all discount rates in the economy down, not just those used to evaluate projects in the public sector. Otherwise, there would be wasteful reallocation of investment from high-yield private projects to low-yield public ones (the opportunity cost principle).

Suppose it is correct that the social rate of discount is below the market, what is the effect on rates of depletion? A lower discount rate will result in slower depletion as the future is given greater weight (than under a higher discount rate), and it will be optimal to save more of the resource for future consumption. Alternatively, market-determined depletion will be too rapid. Another way of thinking about the problem is that the discount rate tilts the time paths of price and extraction, the first up and the second down. The lower the rate, the less the tilt, the closer to equality in consumption across periods. This result, too, will be illustrated in a numerical example.

A newer literature, developed for the choice of an appropriate discount rate for public projects, focuses on discounting in the very long run, with or without uncertainty, as for example becomes relevant when considering problems such as mitigation of climate change. One way of dealing with this is known as hyperbolic discounting, discussed just below. We shall return to the special considerations in analyzing very long run problems in a later section on economic aspects of climate change.

Hyperbolic Discounting

Hyperbolic discounting refers to a discount rate that declines over time (from the perspective of the present). The idea is that it makes a big difference to me if I receive a sum of money in year 2 instead of year 1, but little or no difference (now) between receiving the money (or my descendants receiving it) in year 99 or year 100.

Another way of characterizing hyperbolic discounting is that the discounting relationship between year $n$ and year $n + 1$, as seen from year 1, will change if re-evaluated at year $n - 1$. Hyperbolic discounting in some form is suggested by much survey and experimental research. Early studies in environmental economics include Cropper et al. (1992) and Henderson and Bateman (1995). Theoretical formulations justifying the use of a declining discount rate, in the context of climate change, are given by Arrow et al. (2014). Although hyperbolic discounting has some intuitive appeal, and also some empirical support, it is in my judgment somewhat cumbersome or inelegant, as compared to the standard exponential discounting, and in the early formulations is time-inconsistent.

### 2.6.6 *Perfect Information*

We have implicitly assumed perfect information, in particular about future demand for the resource, and about the size of the stock, for both planner and firms. Suppose either or both is not known; then we have the possible price paths shown in Fig. 2.16

**Fig. 2.16** Depletion under incorrect expectations about future demand or supply

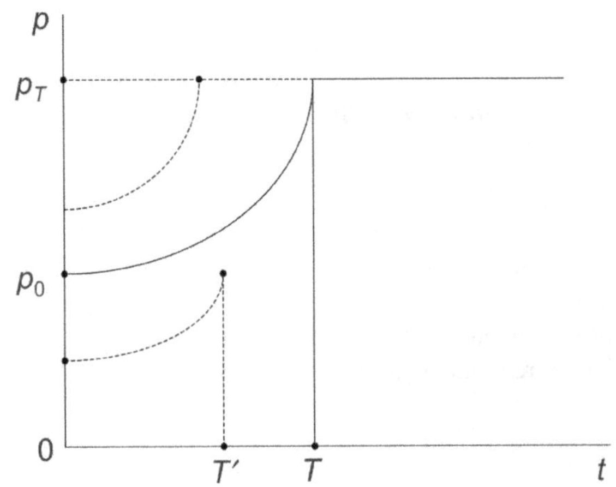

(extending our two-period results to many very short periods to trace out longer, continuous paths).

The "true" path, based on correct expectations about demand and supply, is given by the solid line starting at $p_0$. The flat portion of the line, which begins at time $T$, represents the maximum price, $p_T$, that can be attained, the price of a substitute or the "choke" price, where quantity demanded is zero. At this point, the resource is (optimally) exhausted.

The lower dashed line describes a path characterized by expectations about the size of the stock or the price of a substitute that are too optimistic. Suppose producers believe that the stock is larger than in fact it is. Then, they will produce more early on, with a consequent lower price in each of the early periods, as shown on the Figure. This means that the resource will be exhausted too soon, before $T$, and price will have to jump at $T' < T$, to $p_T$. But a price jump cannot be either optimal or an equilibrium since it would pay to hold some of the resource over at $T'$ for later sale at a higher price.

The higher dashed line describes a path characterized by expectations that are too pessimistic. Suppose producers believe the size of the stock is smaller than it is in fact found to be. Then, they will produce less in each of the early periods, prices will be higher, and some of the resource will be left in the ground when price reaches $p_T$. This cannot be optimal since it would have paid to produce more earlier rather than be left with unused supplies of the resource.

Clearly, expectations are crucial. They need to be correct, and they need to be uniform, i.e., shared by all decision makers. Movement in this direction is provided by futures markets in some resource commodities, such as oil and many metals. These markets, however, extend only modestly into the future, monthly for a couple of years. However: Though efficiency of the market here, perhaps even existence of an equilibrium, becomes problematic, it seems fair to say that a planner or government agency probably has no better idea about future demands or the size of the resource stock. Thus, it's not clear whether lack of foresight leads to a rate of resource depletion that is either too rapid or too slow.

### 2.6.7 Anchoring the Price Path

In Sect. 2.9 below we present a numerical example to show how the price path in a problem of depletion over many periods can be anchored at the correct initial price by exploiting information about the choke price, the size of the stock, and the equation of motion for the price. Here we develop the equations needed to do this in the case of continuous time. For the discrete time analog the process is rather less elegant, but more intuitive, as we shall show.

Given (say linear) demand:

$$p_t = \bar{p} - aq_t$$
$$q_t = \frac{\bar{p} - p_t}{a}$$
$$\text{and } q_t = \frac{\bar{p} - p_o e^{rt}}{a} \quad \text{since } p_t = p_o e^{rt}$$
$$X_0 = \int_0^T q_t dt = \int_0^T \frac{\bar{p} - p_o e^{rt}}{a} dt$$
$$= \frac{\bar{p}}{a} t \Big|_0^T - \frac{p_o e^{rt}}{ar} \Big|_0^T$$
$$= \frac{\bar{p}T}{a} - \left[ \frac{p_o e^{rT}}{ar} - \frac{p_o e^{r*0}}{ar} \right]$$
$$X_0 = \frac{\bar{p}T}{a} - \frac{p_o}{ar}(e^{rT} - 1)$$

We have one equation with two unknowns $p_0$ and $T$. But we also know that the terminal price $p_T = p_0 e^{rT}$ will be the choke price, $\bar{p}$, since production ceases when the demand curve intersects the price axis at $\bar{p}$.

Now we have two equations in two unknowns, and can solve for $p_0$ and $T$ since we know $\bar{p}$, $X_0$, and $a$:

$$(1)\ X_0 = \frac{\bar{p}T}{a} - \frac{p_o}{ar}(e^{rT} - 1)$$
$$(2)\quad \bar{p} = p_o e^{rT}$$

To solve, substitute for $T$ in Eq. (1), from the expression for $T$ in Eq. (2):

$$e^{rT} = \frac{\bar{p}}{p_o} \Rightarrow \ln(e^{rT}) = \ln\left(\frac{\bar{p}}{p_o}\right) \Rightarrow rT = \ln\left(\frac{\bar{p}}{p_o}\right)$$
$$\text{and } T = \frac{\ln\left(\frac{\bar{p}}{p_o}\right)}{r}$$

But note: This solution requires perfect information about supply, i.e., the size of the (industry) stock, $X_0$, as well as the entire (industry) demand curve (and how it behaves over time; we've assumed unchanging). This is a perhaps extreme example of rational expectations. Firms would have to know industry supply and (future) demand and then solve a problem like the one we just solved. Some or all could well get it wrong, so that the industry is off the true price path. Of course if the errors were random, with an aggregate expected value of zero, they would "cancel out," leading to the rational expectations equilibrium.

Finally, there is no guarantee that a planner would do any better. Information about stocks is proprietary, knowledge of demand curves uncertain, and so on - though in the case of the planner, only one agent needs to solve the problem.

## 2.7 Concluding Remarks on Theory

Each of the assumptions required for market-determined depletion of an exhaustible resource to be efficient, or optimal, from a social point of view, is in fact required for efficiency in the allocation of ordinary goods or resources: perfect competition; no externalities; perfect information; and, if the time dimension is important, private and social discount rates the same. Does this mean that there is nothing special, or different, about exhaustible resources, with respect to the basic theorem of welfare economics and the conditions required for it to hold? In my judgment, the answer is "no". Resources are different because the potential market imperfections and failures appear to be empirically more important, as suggested in our discussion. This opens up the very interesting issues of whether and how governments ought to intervene in decisions about resource use, and what policies are best suited to achieve social objectives in each case.

## 2.8 A Note on Empirical Verification of the Basic Model (the Royalty Grows at a Rate Equal to the Rate of Interest)

The obvious way to test the theory would be to look at the behavior of a time series of royalties for particular resources. The obvious difficulty with this approach is that no such series exists; royalties are not readily observable. Moreover, where there are "stock effects", where the remaining resource stock enters either the extraction cost function or the utility function, the time path of the royalty is more complicated and in particular as we have shown, will increase more slowly and may even be falling. One way to proceed in these circumstances would be to look at prices, which are at least observable. In the special case where extraction costs are negligible, it is the price which in theory grows at a rate equal to the rate of interest, the simplest formulation of the "Hotelling Rule". Unfortunately, such cases are at best extremely rare. There is a further problem with using a time series of prices—or royalties—to test the theory, a point to which I'll return later on in this section.

Fortunately, a clever alternative approach has been suggested by Miller and Upton (1985), requiring only information on current prices and costs.

Following Miller and Upton (1985), let the present value of a resource deposit be

$$V = \sum_{t=0}^{T} \frac{(p_t - c)}{(1+r)^t} q_t \tag{2.1}$$

where $c$ is the constant marginal and average cost. Substituting $(p_0 - c)(1+r)^t$ for $(p_t - c)$ in (2.1) we get

$$
\begin{aligned}
V &= \sum_{t=0}^{T} \frac{(p_0 - c)(1+r)^t}{(1+r)^t} q_t \\
&= (p_0 - c) \sum_{t=0}^{T} q_t \\
&= (p_0 - c) S
\end{aligned}
\tag{2.2}
$$

where $S$ is the resource stock, and the unit value is

$$
\frac{V}{S} = p_0 - c \tag{2.3}
$$

So we have obtained an expression for the value of the resource that does not depend on future prices or interest rates. To value a resource, all we need to know is the current selling price and extraction cost. This is the implication of the theory (since the rising royalty is "canceled" by discounting). Miller and Upton test the theory by measuring whether an independent, market indication of the value of oil and gas reserves is consistent with the value calculated as in Eq. (2.3).

To do this, they run a regression of the market value on the difference between current price and extraction cost:

$$
\frac{V}{S} = \alpha + \beta(p_0 - c)
$$

where $\alpha$ and $\beta$ are regression coefficients. If the theory is correct, we would expect $\alpha = 0$ and $\beta = 1$.

In fact, some modification of the theory (see Miller and Upton) to allow for non-constant (decreasing) returns to scale, extraction costs dependent on cumulative production, uncertainty, and taxes, implies that $\alpha < 0$ and $\beta$ is a little less than 1. The intuition is that the royalty (on the right hand side of Eq. (2.2)) is not increasing at rate r, rather more slowly.

For example, extraction costs dependent on cumulative production (or the remaining stock) implies that there is a dividend, in the form of reduced future extraction costs, along with the capital gain associated with holding a unit of the resource in the stock. Since the total return to holding the resource asset is $r$, and part comes as a dividend, the capital gain or increase in the royalty must be less than $r$.

Using data from 39 oil-and-gas-producing firms on 2 or 3 reserve estimates ($S$) over the period 1979–81, share price data ($V$ = share price $\times$ number of shares) for the dates corresponding to reserve estimates, and figures for sale prices ($p$) and operating costs ($c$) from an industry newsletter, Miller and Upton find:

$$
\frac{V}{S} = -2.24 + 0.91(p_0 - c)
$$

just as predicted by the (modified) theory!

Other studies testing the empirical validity of the "Hotelling rule" have been time series estimates of the trend in royalties, calculated as the difference between observed price and (estimated) marginal cost. Results have been mixed:

- Stollery (1983) finds a trend consistent with the theory for the nickel industry, 1952–1973.
- Halvorsen and Smith (1991) reject the theory with data from the Canadian metal mining industry in the aggregate, 1954–74.
- Withagen (1998) mentions this and suggests using disaggregated cost data, but seems unaware of Stollery's article, which does this and fails to reject the theory.

Now let me return to the problem cited at the outset with using time series of prices or, if you can get them, royalties to test the theory. My view is that no study of this type will be reliable because of the many shocks or disturbances outside the model to time series of resource prices; rather, price series are best interpreted as giving evidence on trends in scarcity, which reflects all of the influences, such as new discoveries, technical change, increase or decrease in demand due to macroeconomic fluctuations, changes in market structure (for example, a resource cartel such as OPEC may be able to restrict supply and raise prices for a period of many years, then it may fall apart, then undergo a period of indeterminate effect, and so on), in addition to the (theoretically increasing) royalty. Miller and Upton avoid these confounding influences with their simple indirect test of the theory.

## 2.9 Numerical Examples

### 2.9.1 *Base Case*

Given: demand $q_t = 10 - p_t$ for $t = 0, 1$ as shown in Fig. 2.17, extraction cost $MC_t = 2$, resource constraint$q_0 + q_1 = 10$, and discount rate $r = 0.1$.

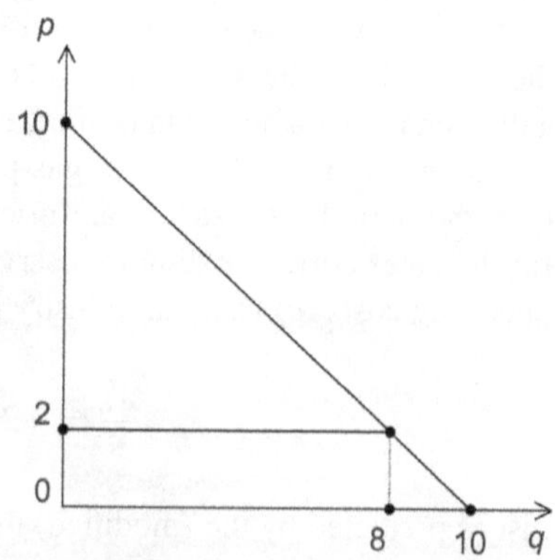


**Fig. 2.17** The demand curve, unchanging over time

Substituting the demand equation into the equilibrium condition, $p_0 - MC_0 = \frac{p_1 - MC_1}{1+r}$, we have

$$10 - q_0 - MC_0 = \frac{10 - q_1 - MC_1}{1+r}$$

Substituting in $MC_t = 2$ and $r = 0.1$,

$$10 - q_0 - 2 = \frac{10 - q_1 - 2}{1.1} \quad \rightarrow \quad 8 - q_0 = \frac{8 - q_1}{1.1}$$

Applying the resource constraint $q_0 + q_1 = 10$,

$$8.8 - 1.1q_0 = 8 - (10 - q_0) \quad \rightarrow \quad q_0 = 5.14$$

$$q_1 = 10 - q_0 = 4.86 \text{ from the stock constraint}$$

$$p_0 = 10 - q_0 = 4.86 \text{ from the demand equation}$$

$$p_1 = 10 - q_1 = 5.14 \text{ from the demand equation}$$

What the solution to this simple numerical problem shows is, as expected, the price rising from one period to the next, to maintain the equality of the present value of the royalty across periods, and the quantity extracted falling, due to discounting (prefer to consume more of the resource early on, when it has a higher present value).

### 2.9.2 Unlimited Stock

Suppose we ignore the stock constraint. Then,

$$p_0 = MC_0 \text{ and } p1 = MC1$$

Substituting in $MC$ and $r$

$$10 - q_0 = 2 \text{ and } 10 - q_1 = 2$$

So $q_0 = 8$ and $q_1 = 8$, but this is impossible since $q_0 + q_1 = 16 > 10$.

### 2.9.3 *No Discounting*

Suppose we ignore discounting. Then,

$$10 - q_0 - 2 = 10 - q_1 - 2$$

Plugging in the stock constraint for $q_1$, we get $q_0 = 5$ and $q1 = 5$. There is an egalitarian allocation of the resource over time or, in other words, price and extraction time paths are flat.

Consider a discount rate larger than zero but smaller than in the base case, i.e. suppose $0 < r < .1$, say, $r = 0.02$. Then,

$$10 - q_0 - 2 = \frac{10 - q_1 - 2}{1.02}$$

$$8 - q_0 = \frac{8 - (10 - q_0)}{1.02}$$

$$q_0 = 5.03, q_1 = 4.97, p_0 = 4.97, p_1 = 5.03$$

So price and extraction paths are flatter than in the case where $r = .1$, but not perfectly flat, as in the case where $r = 0$. In general for $r' < r$, the price rises more slowly and extraction falls more slowly.

### 2.9.4 *Taxes*

Suppose we add a severance tax, = \$1/unit extracted.

$$p_0 - MC_0 - 1 = \frac{p_1 - MC_1 - 1}{1 + r}$$

$$10 - q_0 - MC_0 - 1 = \frac{10 - q_1 - MC_1 - 1}{1 + r}$$

$$7 - q_0 = \frac{7 - (10 - q0)}{1.1}$$

So $q_0 = 5.10, q_1 = 4.90, p_0 = 4.90$, and $p_1 = 5.10$.

Price rises more slowly than without the tax (though the royalty continues to rise at rate $r$), and production is reduced in the present, shifted to the future. In other words, the severance tax leads to conservation of the resource. Why? Because the present value of the tax is reduced by shifting it to the future, and this is accomplished by shifting some production to the future.

This is not a perfectly general result: it depends on the type of tax. For example, suppose that there is a tax on profit per unit, as opposed to per (physical) unit extracted. Let the tax (rate) on profit per unit = $t_0 < t < 1$. Then,

$$(1-t)(p_0 - MC_0) = \frac{(1-t)(p_1 - MC_1)}{1+r}$$

Dividing both sides by $(1-t)$, we have

$$p_0 - MC_0 = \frac{p_1 - MC_1}{1+r}$$

so a profits tax has no effect on the rate of depletion.

But note that the profits tax may affect exploration for new deposits (unless it is imposed on all sectors of the economy, not just the resource sector) since it reduces the return on investment in resource production. This would need to be taken into account in a more general model of depletion that includes exploration.

Finally, consider a tax on gross revenue from resource production—a tax known somewhat confusingly as a royalty. Then,

$$(1-t)p_0 - MC_0 = \frac{(1-t)p_1 - MC_1}{1+r}$$

and $(1-t)$ cannot be canceled from both sides as with the profits tax. To see how the rate of depletion is affected, let $t = .1$ or 10 percent.

$$0.9p_0 - MC_0 = \frac{0.9p_1 - MC_1}{1+r}$$

$$0.9(10 - q_0) - 2 = \frac{0.9(10 - q_1) - 2}{1.1}$$

Proceeding as before, solve for $q_0 = 5.13$, $q_1 = 4.87$, $p_0 = 4.87$, and $p_1 = 5.13$. So the rate of depletion is slowed as some production is shifted to the future. Notice that, if we did divide through by $(1-t)$ above, we would obtain

$$p_0 - \frac{MC_0}{(1-t)} = \frac{p_1 - \frac{MC_1}{(1-t)}}{1+r}$$

Since $(1-t) < 1$, $\frac{MC_0}{(1-t)} > MC_0$ and similarly for $MC_1$. This, in effect, increases the cost of extraction, just as the severance tax does, and the result follows that the present value of the cost is reduced by shifting some production to the future.

### 2.9.5 Expected Demand

Suppose we expect future demand to be greater, say $q_1 = 12 - p_1$. Then

$$p_0 - MC_0 = \frac{p_1 - MC_1}{1 + r} \rightarrow 10 - q_0 - 2 = \frac{12 - q_1 - 2}{1.1}$$

Proceeding as before, solve for $q_0 = 4.19$, $q_1 = 5.81$, $p_0 = 5.81$, $p_1 = 12 - 5.81 = 6.19$. The whole price path is raised as production is shifted to the future to meet the greater demand. The current price, $p_0$, goes up to ration the resource in the current period to conserve it for the future. Does this explain why economists usually oppose price controls in the face of an anticipated shortage?

### 2.9.6 Externalities

Suppose there is an external cost of resource extraction, such as water pollution downstream from the mine. The production decision of a private producer will not be affected (by definition of an external cost), but socially efficient depletion is now determined by

$$p_0 - MC_0 - MEC_0 = \frac{p_1 - MC_1 - MEC_1}{1 + r}$$

where $MEC_t$ is the marginal external cost in period $t$. Suppose $MEC_0 = MEC_1 = 1$. Then, we have

$$(10 - q_0) - 2 - 1 = \frac{(10 - q_1) - 2 - 1}{1.1}$$

and, solving to get $q_0 = 5.10$, $q_1 = 4.90$, $p_0 = 4.90$, $p_1 = 5.10$, we see that some production is shifted from present to future. In other words, in the presence of an externality, the market uses the resource too rapidly. Note: The rate of depletion in this case is the same as in the case of the severance tax though the interpretations are different. Since one way to deal with a pollution externality is to levy a tax equal to marginal damage costs, we could reinterpret the severance tax as a pollution tax which, in this case, would internalize the externality and drive the market-determined rate of depletion to the socially efficient rate.

### 2.9.7 Substitutes

Suppose there is a substitute for the resource, which will be available in the future at a price below the expected future price of the resource. What effect will this have on the price for the resource and the rate of use? Let $p_1 = 4$. Then,

$$p_0 - MC_0 = \frac{p_1 - MC_1}{1+r} = \frac{4-2}{1.1} = 1.82$$

Solving, $p_0 = 1.82 + 2 = 3.82$, $q_0 = 10 - p_0 = 6.18$, $q_1 = 10 - p_1 = 10 - 4 = 6$.

But $6.18 + 6 = 12.18 > 10$. Since there are only 10 units of the resource, and $q_1$ represents use of the resource and substitute, we must have 2.18 units of the substitute produced and just 3.82 units of the resource. The price path drops as production of the resource is shifted to the earlier period.

Interpretation: If we know a lower-priced substitute will be available in the future, we can use the resource more freely early on. If we do not expect to have such a substitute, we would want to conserve.

### 2.9.8 Depletion over Many Periods

We have assumed, for convenience, with no significant loss of generality concerning the cases considered, that the resource will be used over two periods. But, in reality, a resource will be depleted over many periods and an interesting question is: over how many? By the same logic which gives the condition

$$p_0 - MC_0 = \frac{p_1 - MC_1}{1+r}$$

We also have

$$p_1 - MC_1 = \frac{p_2 - MC2}{1+r}$$

and so on. Where does it end? We have a terminal condition: from the demand equation, $p_T = 10$ where $T$ is the terminal period, since this is as high as price can go. At time $T$, the royalty is

$$p_T - MC_T = 10 - 2 = 8$$

Then,

$$p_{T-1} - MC_{T-1} = \frac{p_T - MC_T}{1+r}$$

$$p_{T-1} - 2 = \frac{8}{1.1}$$

So $p_{T-1} = 7.27 + 2 = 9.27$ and $q_{T-1} = 10 - p_{T-1} = 10 - 9.27 = 0.73$. Similarly,

$$p_{T-2} - MC_{T-2} = \frac{p_{T-1} - MC_{T-1}}{1+r}$$

So $p_{T-1} = \frac{9.27-2}{1.1} + 2 = 8.61$ and $q_{T-2} = 10 - p_{T-2} = 10 - 8.61 = 1.39$ and so on. Where does it end? We know that

$$\sum_{t=0}^{T} q_t = 10$$

so when $\sum q_t$ reaches 10, we have solved for the number of periods to exhaustion. In this case, proceeding as before, $q_{T-3} = 1.99$, $q_{T-4} = 2.54$, $q_{T-5} = 3.04$. Note that

$$\sum_{t=T-5}^{T} q_t = 3.04 + 2.54 + 1.99 + 1.39 + 0.73 = 9.69$$

Since $9.69 \approx 10$, we can say that the resource is optimally depleted over a little more than five periods. The approximation is due to the "lumpiness" that accompanies depletion in discrete time periods. We would need continuous time to get an exact solution, as we show in the next section.

### 2.9.9 Depletion over Many Periods: Continuous Time

In Sect. 2.6.7 we presented a method for solving for the optimal stopping time in a continuous time resource extraction model. That example did not include a marginal cost for extraction. The corresponding equations to identify $P_0$ and $T^*$ with a constant marginal cost, $MC$, of extraction are

$$X_0 = \frac{(\bar{p} - MC)T}{a} - \frac{(p_0 - MC)}{ar}(e^{rT} - 1)$$
$$\bar{p} = (p_0 - MC)e^{rT} + MC$$

and the equation of motion governing the price path is $\dot{\mu}/\mu = r$, where in this case $\mu_t = p_t - MC$.

Using the same assumptions as for the previous examples and solving for $T$ and $p_0$ numerically with MATLAB yields results $T = 5.45$ and $p_0 = 6.64$. The full 10 units of the resource will be optimally consumed in 5.45 periods, which is consistent with the discrete time example where almost all the resource was consumed in five periods and all in just over five periods. The corresponding price path is shown in Fig. 2.18.

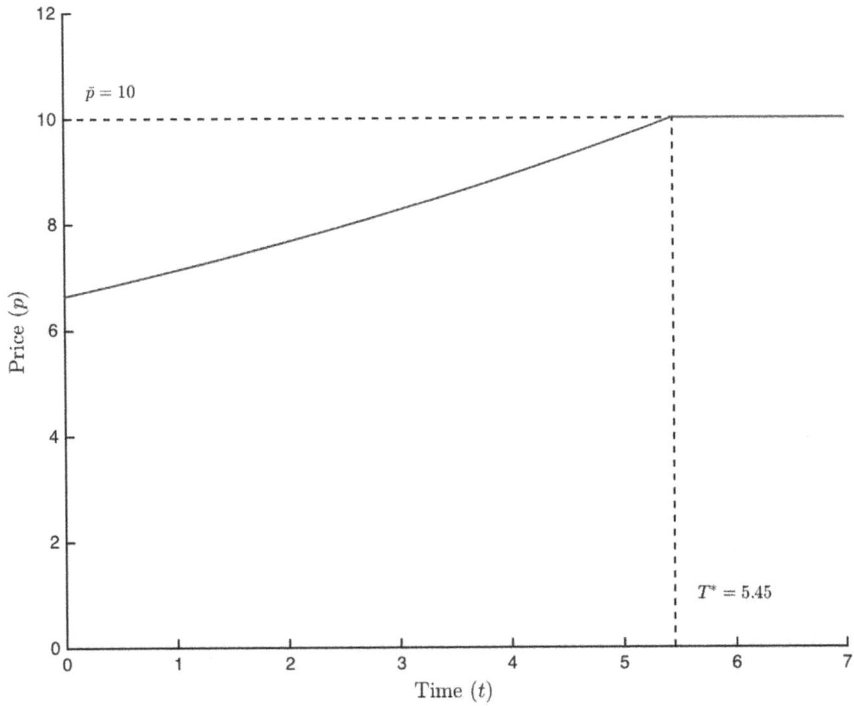


**Fig. 2.18** The optimal price path in continuous time. The stock is exhausted at $T = 5.45$

## *2.9.10 Incorrectly Anticipated Stock*

If the resource manager has an incorrect expectation about the initial stock, the corresponding price path will be sub-optimal. Figure 2.19 shows sub-optimal price trajectories in addition to the optimal price trajectory generated with correct expectations about the initial stock (blue path).

When the expected stock is too low ($\mathbb{E}[X_0] = 8$), the initial price is too high ($p_0 = 6.93$), the stopping time is too soon ($T_{\mathbf{E}[X_0]=8} = 4.83$), and only 8 units of the stock are extracted before the choke price is reached (red path). This is clearly not optimal since some of the resource is left in the ground. When the expected stock is too high ($\mathbb{E}[X_0] = 12$), the initial price is too low ($p_0 = 6.38$), the stopping time is too soon ($T_{\mathbf{E}[X_0]=12} = 3.7$), and although the stock is consumed completely, there is a price jump when the stock is exhausted, which, as we have argued earlier, cannot be optimal (green path).

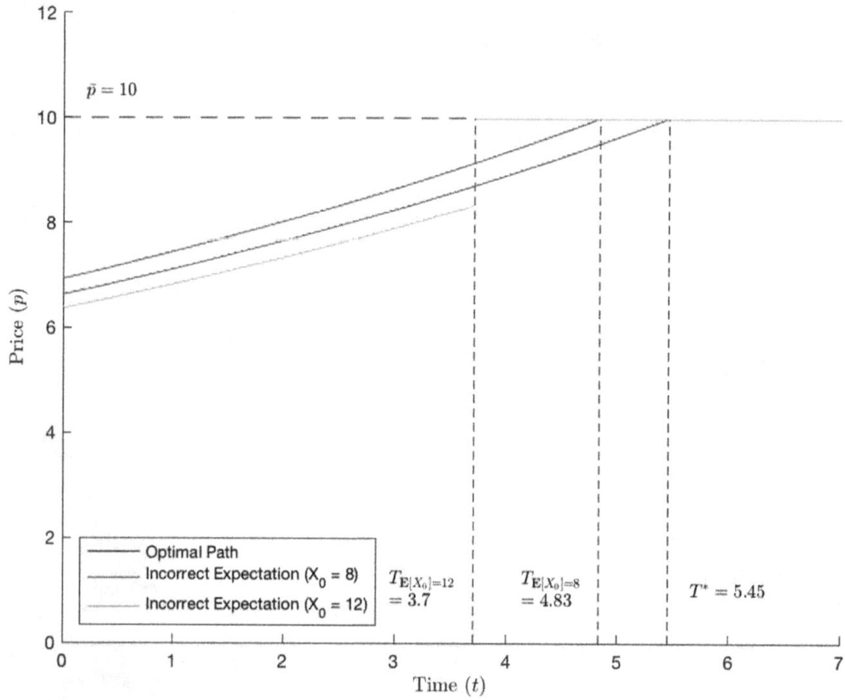


**Fig. 2.19** Price paths with incorrect beliefs about the initial stock

### *2.9.11 Incorrectly Anticipated Choke Price*

Similarly to the previous example, if the resource manager incorrectly anticipates the choke price, then the corresponding price path will be sub-optimal. Figure 2.20 shows sub-optimal price trajectories in addition to the optimal price trajectory generated with correct expectations about the choke price (blue path).

When the expected choke price is too low ($\mathbb{E}[\bar{p}] = 8$), the initial price is too low ($p_0 = 5.17$), the stopping time is too soon ($T_{\mathbf{E}[\bar{p}]=8} = 2.25$), and although the stock is consumed completely, there is a sub-optimal price jump when the stock is exhausted (red path). When the expected choke price is too high ($\mathbb{E}[\bar{p}] = 12$), the initial price is too high ($p_0 = 8.17$), the stopping time is too soon ($T_{\mathbf{E}[\bar{p}]=12} = 2.6$), and only 2.48 units of the stock are extracted, with the rest left in the ground (green path).

In this example, the stopping time is more sensitive to incorrect expectations about the choke price than the initial stock.

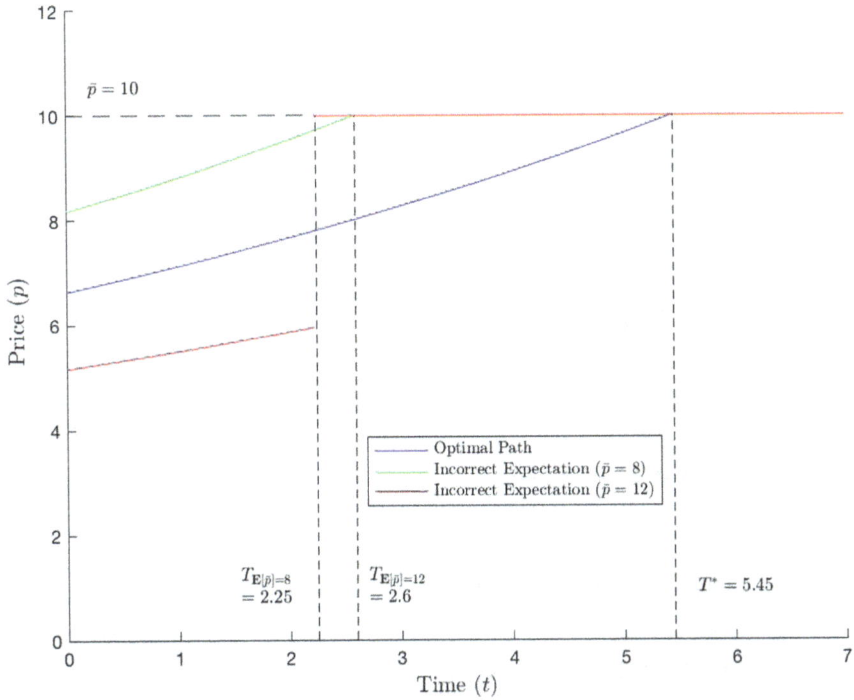


**Fig. 2.20** Price paths with incorrect beliefs about the choke price

## 2.10 Some Last Notes on Scarcity Rents

It is important to note the similarities between Ricardian rents and Hotelling royalties but also the differences. In both cases, the value of the resource is determined by its scarcity and the availability or scarcity of alternatives, that is, the resource's supply relative to demand and the availability of substitutes. Both the rent and royalty are calculated as the difference between price and marginal cost, as shown in Fig. 2.21. However, the two concepts measure somewhat different things. Ricardian rents typically describe a heterogeneous fixed resource, like land or capital, in a single period of time. Price is set by the unit on the margin, and all the other units in operation receive a rent, that is, an economic profit, based on the extent to which that unit's marginal costs are lower than the marginal cost of the unit that is on the margin. Hotelling royalties on the other hand describe a consumable, extractible resource, consumed over many periods. In the next section we relate the royalty to the cost of what has been called a "backstop," or substitute for the resource, which comes into play when the resource is exhausted. This formulation in turn allows us to interpret the royalty as a kind of dynamic rent.

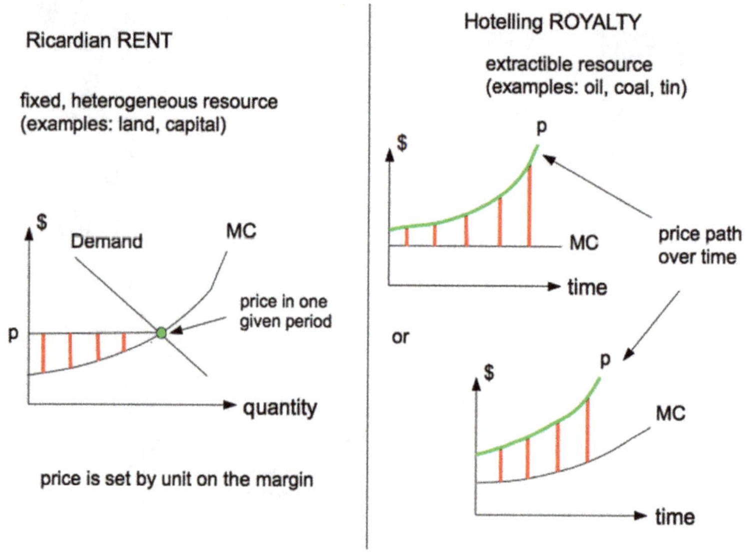


Fig. 2.21 Contrasting Ricardian rents and Hotelling royalties

## 2.11 Backstop Technologies

Consider an exhaustible resource and a substitute or "backstop" technology that is more expensive but available in virtually unlimited quantities. Examples in the energy sector would be exhaustible fossil fuel and unlimited solar.

The cheap resource is used first (push the higher costs into the future, where they are more heavily discounted). Along an optimal path,

$$p_t - c = (p_0 - c)(1 + r)^t$$

where c = assumed constant marginal and average cost of production of the resource.

At the switch date T to the backstop,

$$p_T = c + (p_0 - c)(1 + r)^T$$

Since the backstop is by definition not a scarce resource, there is no scarcity rent or royalty attached to its production, so price = marginal cost, $c_b$.

The switch to the backstop occurs when $p_T = c_b$ so,

$$c_b = c + (p_0 - c)(1 + r)^T$$

Rewriting,

$$p_0 - c = \frac{c_b - c}{(1 + r)^T}$$

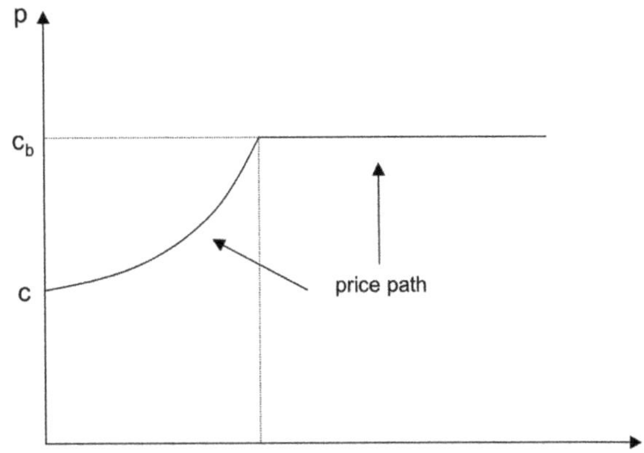


**Fig. 2.22** Price path up to backstop technology with marginal cost $c_b$

The royalty at $t = 0$, $p_0 - c$, is the difference between $c_b$ and $c$, discounted back from the switch date $T$. Note the similarity to Ricardian rent, the UNDISCOUNTED difference in costs. Here the resource and substitute are used in sequence, not simultaneously, as in the Ricardian model (Fig. 2.22).

Substituting this expression for the royalty into the equation for the time-path price,

$$p_t = c + \frac{(c_b - c)(1+r)^t}{(1+r)^T} = c + \frac{c_b - c}{(1+r)^{T-t}}$$

At t = 0,

$$p_0 = c + \frac{c_b - c}{(1+r)^T},$$

and for c = 0 (negligible cost of extraction),

$$p_0 = \frac{c_b}{(1+r)^T}.$$

The price varies positively with the cost of the backstop, and negatively with the discount rate and the switch date. Note the nonlinear way in which changes in demand or in reserves can affect the price, because of the way in which the switch date enters the expression for the royalty.

For fixed demand of $D$ units per year and known reserves of $R$ units, $T = R/D$.

Now suppose estimated reserves $R$ are doubled or annual demand $D$ is halved. Then $T$ is doubled.[8]

Suppose the initial price is just 10% of backstop cost, that is, $p_0 = 0.1c_b$. Then if $T$ is doubled,

$$(1+r)^{-T} = 0.1$$

[8]Note: this is not unrealistic. World oil reserves have gone from 70 billion bbl in 1950 to 1000 billion bbl in 2000. Japan uses approximately 1/2 the energy per capita used in the U.S.

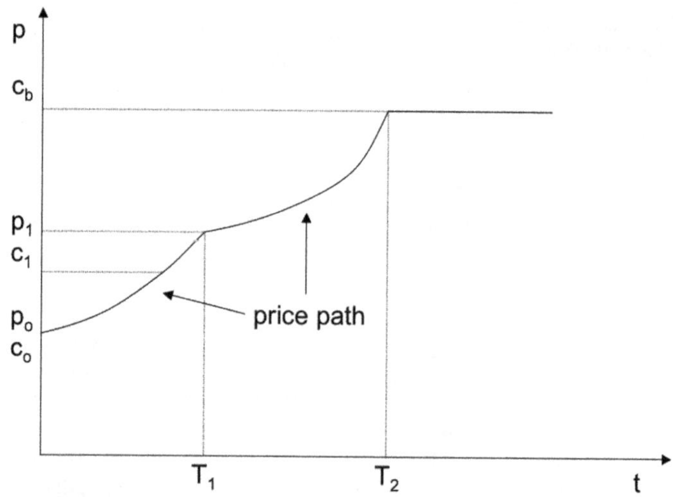


**Fig. 2.23** Price path with multiple substitutes

$$(1+r)^{-2T} = (1+r)^{-T}(1+r)^{-T} = 0.1^2 = 0.01$$

so the initial price becomes $p_0 = 0.01c_b$, just 1% of the backstop cost. This may be one reason for large fluctuations in resource prices—sensitivity to "offstage" developments bearing on the switch date.

Now suppose there are in fact many substitutes for the free or low-cost resource along the path to the ultimate, unlimited backstop. This is of course more realistic, and is a feature of the Nordhaus model for the allocation of energy resources over time, discussed in the next section. First, we extend the simple model to consider many substitutes by adding in just one (plus the backstop), with no loss in generality. The resulting price path is shown in Fig. 2.23.

Proceeding as before, the royalty on the substitute is the discounted difference in costs

$$\frac{c_b - c_1}{(1+r)^{T_2-t}} \qquad T_1 \le t \le T_2$$

At $t = T_1$,

$$\text{royalty} = \frac{c_b - c_1}{(1+r)^{T_2-T_1}}$$

This gives us an expression for the price, $p_1$, at $T_1$, the switch date from the resource to the substitute:

$$p_1 = c_1 + \frac{c_b - c_1}{(1+r)^{T_2-T_1}}$$

Note that $p_1 > c_1$, since the substitute is scarce, and commands a royalty. The royalty on the resource is

$$\frac{p_1 - c_0}{(1+r)^{T_1-t}} \qquad 0 \leq t \leq T_1$$

At $t = 0$,

$$p_0 = c_0 + \frac{p_1 - c_0}{(1+r)^{T_1}}$$

where

$$p_1 = c_1 + \frac{c_b - c_1}{(1+r)^{T_2-T_1}}$$

So the cost of the backstop casts a long shadow, all the way back to the initial price of the resource. Of course, if $T_2$ is far in the future, or r is high, this is less relevant.

Note that the royalty rises on the intensive margin, at rate $r$, and then falls on the extensive margin. This is a discrete version of the behavior of the royalty when cost is a function of cumulative extraction or the remaining stock, as in our optimal control model (Sect. 2.5.1), where the royalty can fall all the way to zero in the transition to "average rock," a very low grade of the resource which is available in abundance, with correspondingly little scarcity value.

### *2.11.1 Nordhaus Energy Model and Beyond*

This simple model underlies one of the earliest and most influential analyses of the world energy economy, the Nordhaus (1973a) model for the allocation of energy resources over space and time. Of course this model presents a far more detailed and realistic picture of the transition from one energy source to another along the path to the backstop, with empirical content regarding comparative costs, broken down into the separate components of extraction, transportation, and processing, on a regional basis. On the other hand, the problem is somewhat "simpler", in that it involves cost minimization subject to a given pattern of end uses rather than maximization of profits or consumer surpluses, both of which require knowledge of the shape of demand curves and their behavior over time.

At this point a question arises as to why it makes sense to look in any detail at a model developed over 40 years ago, given developments in energy technologies, markets and policies since then. One answer is of course that it illustrates how our simple model can be used, suitably expanded yet still readily recognizable, to shed light on a real-world problem. But beyond this is its continuing influence, as evidenced in a recent (2015) symposium on the model in the Journal of Natural Resources Policy Research.

Formally, the problem posed is to sequence exhaustible resources and substitutes "optimally" to minimize the present value of the costs of meeting a set of exogenous regional demands or end uses over time. The solution is generated by a linear programming model, which yields time paths of royalties and prices along with the sequencing of energy sources. If one wants to make the connection between an optimal allocation and a competitive equilibrium, then model solutions can be used to shed light on the evolution of energy sources and prices in a competitive industry over time, starting in 1970. Alternatively, looking back in time, the differences between calculated prices and quantities from the model and those observed can be interpreted as a measure of the deviation of actual outcomes from an efficient configuration. More recent programming/planning models, which we discuss briefly at the end of this section, are typically devoted to policy issues faced by a relatively small, well-defined region and play out over a shorter planning period (the Nordhaus model is global in scope and generates results over four 50-year periods) and are still more detailed and realistic, but the analytical structure is not significantly different.

An alternative approach, taken in a number of energy/economy models over the past two decades, puts more emphasis on environmental constraints, in particular those imposed by climate change. This changes the focus from the amounts of a resource such as oil or coal in the ground (or under the sea) to the impact on atmospheric concentrations of carbon dioxide and other greenhouse gases of continued production. In this setting it may and probably will be optimal to leave some of the resource in the ground, even if the conventional costs of production are not rising. Here too some of the earliest and in my judgment most influential work is by Nordhaus, whose DICE (Dynamic Integrated Climate/Economy) model is discussed in a later section on climate as the ultimate resource (Nordhaus, 1993, plus more recent iterations). With perhaps a very few exceptions, the climate/economy models, including DICE, however lose the detail of the transition from one energy source to another, focusing instead on the broad relationships between output, emissions and concentrations.

In symbols, the objective is to determine the allocation of energy resources over time that minimizes the cost of meeting a set of given demands over time.

$$\min_{i,j} \sum_{i,j,k,l,m} \frac{c(i,j,k,l,m)\, x(i,j,k,l,m)}{(1+r)^m}$$

where $c$ is unit production cost and $x$ is the flow of resource $j$ from region $i$ to demand category $l$ in region $k$ at time $m$, measured in delivered thermal content (btu) of final product. The model is disaggregated into four regions producing the resource ($i =$ US, WE, ME, ROW), 17 types of resource ($j =$ oil, shale, natural gas, breeder, etc.), four regions consuming the resource ($k =$US, WE, J, ROW), five demand sectors ($l =$ electricity, industrial non-electric, residential non-electric, substitutable transport, non-substitutable transport (mostly air)), and nine time periods ($m =$ four 50-year periods, beginning in 1970, dividing the first into five decades).

Demand in each sector or category is exogenous, but not constant; a fixed path, based on past trends (though note these reflected low and falling or steady energy prices until shortly before publication). The different energy sources are assumed substitutable (except for non-substitutable transport). Nordhaus claims that fairly stringent environmental standards are incorporated either as constraints or as part of costs, though clearly standards have grown more stringent since 1973 and involve additional threats to the environment such as emissions of greenhouse gases like carbon dioxide, so called because they trap heat in the atmosphere.

Unit production costs are broken down into three components: extraction costs $ex$, transport costs $tr$, and processing costs $pr$.

$$c(i, j, k, l, m) = \frac{ex(i, j) + tr(i, j, k) + pr(j, l)}{(1+r)^m}$$

There are of course constraints on the cost minimization. For each region-resource pair $i$, $j$ the supply of resource $j$ from region $i$ must be at least as great as activity over all demands, regions, and periods that use $i$, $j$:

$$R(i, j)_{ME,oil} \geq \sum_{k,l,m} x(i, j, k, l, m)/\sigma$$

where $\sigma$ is a conversion factor for thermal efficiency (say btu/bbl).

For each region, end-use, period triple, $k$, $l$, $m$, demand in region $k$ for end-use $l$ in period $m$ must be no greater than can be sustained by the flow of all resources $j$ from all regions $i$ for purpose $l$:

$$D(k, l, m)_{US,elec,2010} \leq \sum_{i,j} x(i, j, k, l, m)$$

Other constraints, for example on amount of imported oil, or to protect the environment, could in principle be added.

In greatly aggregated form here (but see Nordhaus (1973a)), results for the US are:

- 1970–80: Domestic oil and gas (cheapest, extraction almost free, transport cost low), plus some imported oil
- 1980–2000: Imported oil plus imported LNG, so heavy dependence on foreign sources
- 2000–2020: Imported oil and LNG, high-cost domestic oil and gas, transition to coal and shale, introduction of nuclear fission for electricity sector
- 2020–2070: Coal, gasified and liquefied coal, shale, and introduction of breeder

We should note a few caveats to these results. Due to linearity, the model produces "bang-bang" solutions: each region uses up all of one resource before moving on to the next. This effect is seen because in the model costs don't rise as production is expanded. The simplification is unrealistic but could be handled in the linear programming format by capacity constraints, or by a move to, say, quadratic programming: minimizing a quadratic cost function subject to linear constraints. Furthermore, the model doesn't allow for historical developments, such as government development and subsidy of nuclear fission, or OPEC-induced oil prices above efficiency prices. But of course this is one of the purposes of the model: to generate an "efficient" solution, against which historical developments can be compared. Perhaps most importantly, it does not give what would today be considered sufficient attention to the environmental disruption and hence the external costs that accompany production and use of energy especially from fossil fuels. Climate change is of course receiving the most attention today, but starting in the 1970s, with passage of the federal Clean Air and Clean Water Acts and subsequent amendments, policy at all levels of government has made environmental impacts a key issue in the consideration of existing and planned energy developments.

The programming solution yields a set of shadow prices (attached to the resource constraints), the royalties. As in our simple model, the royalty on each resource increases at a rate equal to the rate of interest along the path to the next in the sequence. The royalties, when added to the costs, give a price path. In 1970, the efficiency price of oil as calculated in the model was $1.20 per barrel (in 1970 dollars!), not too different from then-current prices (which were however about to increase dramatically due to actions by OPEC, the oil producers' cartel). The model calculates that the price of oil, in a perfectly competitive global market, would rise to $7.12 in 2010, and presumably a little higher currently (2016). Note that $7.12 in 1970 dollars translates to about $30–35 in current dollars, reasonably close to current—though fluctuating—prices in the world oil market, a market characterized by a failure of the cartel to agree on the allocation of production cuts that would be needed to sustain a higher price.

These calculations, although they put the model numbers in perspective, are however somewhat misleading, as they suggest that current prices, which clearly depend on recent developments in the global economy, technical advances in extracting and producing oil and natural gas ("fracking"), and so on, have been accurately "predicted" far into the future (from the perspective of 1973). The model is better understood as shedding light on the question of how to think about the planning of energy production and use involving multiple sources and regions over long periods of time, a question given urgency at the time of publication by the turmoil in the world oil market.

As we noted at the outset of our discussion of the Nordhaus model, more recent programming/planning models are typically devoted to analyzing a range of issues involving, in addition to identifying least-cost energy scenarios, including responses to restrictions on emissions, especially of greenhouse gases, evaluating new technologies and the effects of regulations, taxes and subsidies more generally. These and still other issues are typically studied for relatively small and well-defined regions or countries using MARKAL (for MARKet ALlocation), the main analytical tool, a linear programming model involving over 10,000 equations and constraints, which can be specified as desired for a particular region. The model was originally sponsored by the U.S. Department of Energy and the International Energy Agency (IEA) and developed at Brookhaven National Laboratory for energy system modeling and analysis in the late 1970s, and has been widely used since in more than 40 countries to analyze a broad range of issues in energy planning and environmental policy formulation. More recently a new and improved version of MARKAL, the TIMES model, has been developed by the IEA, although the code for MARKAL will continue to be supported so it remains an option. There is to my knowledge not much of an academic literature on the models, but more on the features of each can be found online.

# Chapter 3
# Renewable Resources

**Abstract** Turning to renewable resources, the new feature is some process for regeneration, accomplished by introducing a biological growth function, the widely applicable logistic law, into the optimal control model for exhaustible resources. Although a quite different result might be expected, and this is in fact the case, it turns out that the optimal management regime can be characterized in a way that also centrally involves the rate of interest. Depending on assumptions about the behavior of the growth function and the interest rate, a variety of outcomes are derived. In particular, under certain conditions the bio-economic optimum can be shown to be equivalent to what might be called a purely biological optimum, the maximum sustainable yield. This analysis has been for a generic renewable resource. The chapter also develops a model for the optimal harvesting of a particular resource, timber, or more generally forest management. This involves no mathematics beyond the elementary calculus used in the modeling of optimal extraction of an exhaustible resource, though now in a setting of continuous time and, following the conventional treatment, the decision variable is the optimal harvest date or rotation period rather than the amount of timber to be harvested at each point in time. Four cases are considered: the optimal harvest date for a one-shot harvest (which might be appropriate for a very slow-growing species); the optimal rotation period where repeated harvests are indicated, the conventional approach; a purely biological model, in which the objective is to maximize not value but the sustainable physical yield, which is contrasted with the bioeconomic model as in the case of a generic renewable resource; and the same analyses (though it is sufficient to look at how results are affected in the first case, the one-shot harvest) taking into account that a standing forest can also provide value. This leads quite naturally to the next chapter, on environmental dynamics.

Turning from exhaustible to renewable resources, the new feature is clearly some process for renewal, or regeneration, of the resource. A straightforward way to model this is to introduce a biological growth function into a model of optimal exhaustible resource use. Looking ahead to results, we might expect something very different from those in the exhaustible resource model, since a renewable resource will normally be managed sustainably, that is, with repeated harvests from a maintained stock. This is in fact the case, but it turns out that the optimal management regime





A. C. Fisher, *Lecture Notes on Resource and Environmental Economics*,
The Economics of Non-Market Goods and Resources 16,
https://doi.org/10.1007/978-3-030-48958-8_3

can be characterized in a way that has the same "flavor" as in the exhaustible resource model, in that it also centrally involves the interest rate.

We first develop an application to a particular renewable resource, a model of optimal timber harvesting, or more generally, forest management. This requires no mathematics beyond the elementary calculus used in the modeling of optimal extraction of an exhaustible resource over two (or more) discrete time periods. However, we do need to move from discrete to continuous time, as this is the standard format for analysis of the timber harvesting problem and other renewable resource problems. The next section provides a brief derivation of the transition from discrete to continuous time.

Alternatively, or in addition, we might look at another renewable resource, fisheries. I won't go there, for two reasons. One, I have not worked in this area, and have not given any lectures, and two, it's my impression from a casual acquaintance with the literature that the main concern has been with the common property nature of the resource and policy alternatives for getting around this, such as individually transferable quotas (ITQs). Of course dynamics can come into play, but these will be addressed in detail in the sections on timber harvesting and generic renewable resources. A feature of the fisheries problem not addressed there is the possibility of extinction of the resource, though note this is also of particular relevance due to the common property nature of the resource. A comprehensive and accessible treatment of fisheries economics for those interested in the special features of this resource is given by Karp (2017).

Following the discussion of timber harvesting and forest management we return to the more general problem of optimal use of renewable resources by introducing a biological growth function, as in the timber harvesting problem the widely applicable logistic law, into the optimal control model for exhaustible resources. We consider a generic renewable resource, with no special features appropriate to any particular resource such as a fishery or a forest, but this still yields some interesting and reasonably general results, which also correspond to those obtained in the forestry problem.

## 3.1 Continuous Time

Table 3.1 shows the difference made by continuous compounding of an initial sum, here 100, as opposed to discrete or annual compounding.

$$e \text{ is defined as } e^r = \lim_{n\to\infty}\left(1+\frac{r}{n}\right)^n \qquad e^1 \approx 2.7182818$$

Important: we're taking the limit here of the *time step*.

The value $V$ of an amount $P$ after $T$ years is:

$$V = (1+r)^T P \qquad\qquad V = e^{rT} P$$

**Table 3.1** Continuous compounding, assume $r = 10\%$, $x = 100$, $T = 10$

| | Value end of one year | | Value end of $T$ years | |
|---|---|---|---|---|
| Annual | $(1+r)x$ | 110 | $(1+r)^T x$ | 259.37 |
| Twice a year | $(1+\frac{r}{2})^2 x$ | 110.25 | $(1+\frac{r}{2})^{2T} x$ | 265.33 |
| Every trimester | $(1+\frac{r}{3})^3 x$ | 110.337 | $(1+\frac{r}{3})^{3T} x$ | 267.43 |
| Every quarter | $(1+\frac{r}{4})^4 x$ | 110.381 | $(1+\frac{r}{4})^{4T} x$ | 268.51 |
| Weekly | $(1+\frac{r}{52})^{52} x$ | 110.506 | $(1+\frac{r}{52})^{52T} x$ | 271.57 |
| Continuously | $e^r x$ | 110.51709 | $e^{rT} x$ | 271.83 |

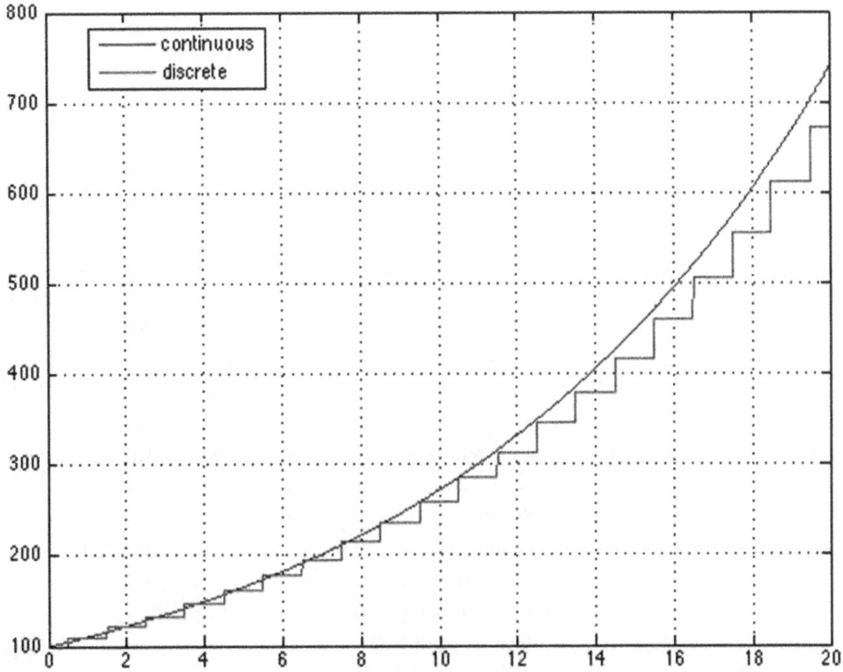


**Fig. 3.1** Discrete versus continuous time: $100(1+r)^t$ versus $100e^{rt}$ for $r = 0.1, t \in [1, 20]$

Figure 3.1 displays these results, expanded to include values for years between 1 and 20.

Now, let's reverse the process and ask, how much money, $P$, would I need to set aside today to have an amount $V$ available $T$ years from now:

$$P = \frac{1}{(1+r)^T} V \qquad\qquad P = e^{-rT} V$$

In other words, $P$ is the present value of receiving/spending $V$ in the future.
In discrete time we set up problems as:

$$\max \sum_{t=0}^{\infty} \frac{1}{(1+r)^t} x(t)$$

In continuous time we set up problems as:

$$\max \int_0^{\infty} e^{-rt} x(t)\, dt$$

## 3.2 Analytics of Optimal Timber Harvesting

In the case of exhaustible resources we asked how much quantity to extract in each period. We can ask the same question with renewable resources; for example, how much to fish each year given the fish population's regeneration dynamics. For renewable forestry resources, instead of asking how much timber to extract each period we can ask, following the conventional treatment, how long to wait before harvesting a stand of trees or, in the case of multiple harvests, what is the optimal rotation length.

We consider four cases, of increasing complexity and perhaps realism, elaborating on a model presented by Hyde (1980). The first is: when do we harvest a stand of (assumed even-aged) trees, if there is to be just a single or one-shot harvest? Although this ignores potential future harvests it might be appropriate for example in the case of a very slow-growing species, for which the discounted value of future harvests is very low. The second case involves repeated harvests and determination of the optimal rotation period, the time to the first harvest and between successive future harvests. This is the way the problem of optimal harvesting is ordinarily stated, presumably because it is applicable to a wide variety of commercial timber resources. Interestingly, the solution was first demonstrated by a German forester, Martin Faustmann (1849) and in consequence is sometimes referred to as the Faustmann rule.[1]

The third case contrasts the results of a bio-economic model, which includes elements of both biology (a natural growth law, such as the logistic, which we shall presently describe) and economics (maximizing present value), with an approach sometimes advocated by resource managers, a purely biological model in which the objective is to maximize not value, but the sustainable physical yield of the resource.

In all three cases, the forest is considered solely as an asset yielding an extractive resource: timber. But it is increasingly recognized that a standing forest can also provide value. We have seen the implied conflict play out in recent years for example in the debate about whether to harvest stands of old-growth redwood in Northern California. Our fourth case extends the bio-economic model to consider how the optimal harvest date or rotation period is affected if the standing forest also yields value. In fact, as we shall see, depending on the initial values of the competing forest uses and the way these values change over time, the time to harvest may be never.

[1]The original article was published in German. An English translation was reprinted in *Journal of Forest Economics* in 1995.

### 3.2.1 One-Shot Harvest

The problem is to maximize V, the present value of the harvest. In symbols, this is

$$\max_T V = pQ(T)e^{-rT}$$

where $p$ is the price of timber net of harvest cost, $Q$ is the volume of wood, $T$ is the harvest date, and $r$ is the discount rate.

The annual growth in biomass, $\frac{dQ}{dT}$, as for a renewable resource generally, can be represented by the logistic law

$$\frac{dQ}{dT} = \gamma Q\left(1 - \frac{Q}{K}\right) = \gamma Q - \frac{\gamma}{K}Q^2$$

where $K$ represents a carrying capacity, a maximum tree or forest size, expressed in the same units as $Q$, the volume of wood, and $\gamma$ is the "intrinsic" growth rate of the resource, i.e. the rate of exponential growth before crowding sets in. The evolution of $Q$ then follows a trajectory like the one depicted in Fig. 3.2: as $Q$ increases, $\frac{dQ}{dT}$ first increases then decreases, because $\frac{\gamma}{K} \ll \gamma$. For $Q$ close to zero,

$$\frac{dQ}{dT} \approx \gamma Q \qquad \text{or} \frac{\frac{dQ}{dT}}{Q} \approx \gamma$$

As $Q$ approaches its limit of the environmental carrying capacity $K$,

$$\frac{dQ}{dT} \approx 0$$

Taking natural logs of the objective function we're optimizing,

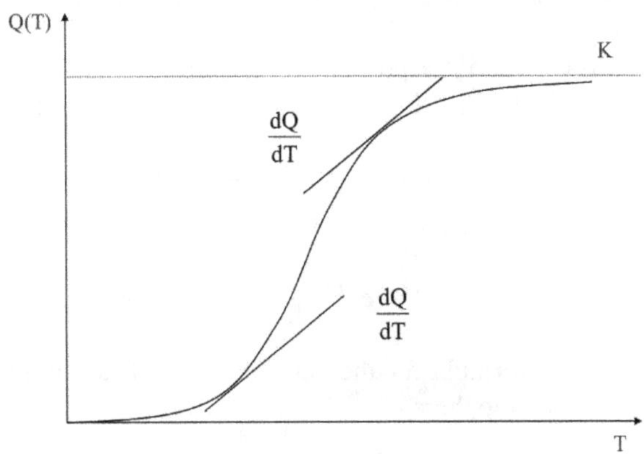


**Fig. 3.2** Volume of biomass $x(T)$ first increases rapidly (exponential growth) then at a decreasing rate as a stand of trees ages

$$\ln V = \ln p + \ln Q(T) - rT$$

The first order condition is

$$\frac{dlnV}{dT} = \frac{1}{V}\frac{dV}{dT} = 0 + \frac{1}{Q}\frac{dQ}{dT} - r$$

(assuming $p$ is not changing over time)

$$\frac{dV}{dT} = V\left(\frac{1}{Q}\frac{dQ}{dT} - r\right) = 0$$

$$\frac{dQ}{dT}\bigg/ Q = r$$

This result has a straightforward interpretation, involving the opportunity cost of keeping money tied up in trees. Before the optimal harvest date, $T*$, the resource asset grows at a rate higher than the rate of return in the economy, the interest rate, so it pays to leave the trees in the ground, or in other words to invest in trees. After $T*$, when the asset is growing more slowly (at a rate less than the rate of interest) it would have been better (more profitable) to harvest just at the point where the rate of growth falls to the rate of interest and invest the proceeds of the timber sale elsewhere in the economy. Beyond that point, the resource in the ground is yielding lower returns than an alternative investment.

### 3.2.2 *Repeated Harvests*

It will normally be appropriate to take into account the potential value of future harvests, net of replanting costs, although the one-shot model may be sufficient for particularly slow-growing species, given discounting. The problem can be stated as follows:

Maximize V, the present value of repeated harvests, taking into account replanting cost $c$, by finding the optimal rotation period $T$ Solution:

$$\max_{T} V = [pQ(T) - c]\, e^{-rT}\left(1 + e^{-rT} + \cdots + e^{-nrT}\right)$$

Note:

$$1 + e^{-rT} + \cdots + e^{-nrT} + \cdots = \left(1 - e^{-rT}\right)^{-1}$$

from the formula for the sum of terms of an infinite geometric series where the common ratio, here $e^{-rT}$, is less than one.

$$V = (pQ - c)\, e^{-rT}\left(1 - e^{-rT}\right)^{-1}$$

$$\ln V = \ln(pQ - c) - rT - \ln\left(1 - e^{-rT}\right)$$

$$\frac{\ln V}{dT} = \frac{d\ln(pQ - c)}{dT} - r - \frac{d\ln(1 - e^{-rT})}{dT}$$

$$\frac{1}{V}\frac{dV}{dT} = \frac{1}{pQ - c}\, p\, \frac{dQ}{dT} - r - \frac{1}{1 - e^{-rT}}\, re^{-rT}$$

$$\frac{dV}{dT} = V\left[\frac{p\frac{dQ}{dT}}{pQ - c} - r - \frac{re^{-rT}}{1 - e^{-rT}}\right] = 0$$

$$\frac{p\frac{dQ}{dT}}{pQ - c} = r + \frac{re^{-rT}}{1 - e^{-rT}} = \frac{r\left(1 - e^{-rT}\right)}{1 - e^{-rT}} + \frac{re^{-rT}}{1 - e^{-rT}} = \frac{r}{1 - e^{-rT}}$$

If $c$ is small relative to $pQ$ then

$$\frac{p\frac{dQ}{dT}}{pQ - c} \simeq \frac{p\frac{dQ}{dT}}{pQ} = \frac{\frac{dQ}{dT}}{Q}$$

$$\frac{dQ}{dT}\Big/ Q \simeq \frac{r}{1 - e^{-rT}}$$

$$\text{As } T \to 0,\ e^{-rT} \to 1 \text{ and } \frac{r}{1 - e^{-rT}} \to \infty$$

$$\text{As } T \to \infty,\ e^{-rT} \to 0 \text{ and } \frac{r}{1 - e^{-rT}} \to r$$

The behavior of $\frac{r}{1 - e^{-rT}}$ is shown on Fig. 3.3, along with the implied optimal dates for the one shot harvest and the first of the repeated harvests. Note that the repeated harvest comes sooner. This result too has an economic interpretation. With future harvests now relevant, the forest manager faces an additional opportunity cost of leaving trees in the ground: the more rapid growth of a newly planted stand, as illustrated in Fig. 3.2.

### *3.2.3 A Numerical Example Comparing One-Shot and Repeated Harvesting*

We develop a numerical example comparing the optimal rotation length for one-shot and repeated harvest problems when growth follows the logistic growth law.

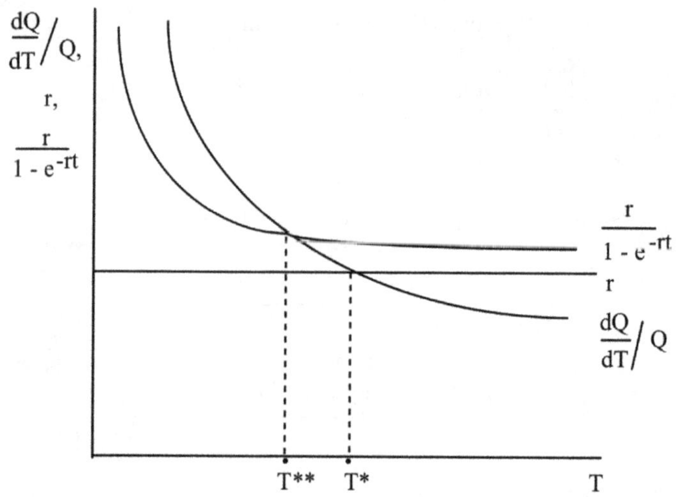


**Fig. 3.3** Considering future harvests shortens the time to harvest ($T^{**} < T^{*}$)

The logistic growth equation is given below in Eq. 3.1. It is a function of two parameters: $\gamma$, the growth rate; and $K$, the maximum stock size.

$$\frac{dQ}{dt} = \gamma Q(1 - \frac{Q}{K}) \tag{3.1}$$

When integrated, the logistic growth function gives Eq. 3.2 as the total stock at time $t$, when starting from an initial stock of $Q_0$.

$$Q(t) = \frac{K Q_0}{Q_0 + (K - Q_0)e^{-\gamma t}} \tag{3.2}$$

For this example we use parameter values[2] of $\gamma = 0.4669$ and $K = 93380$. We assumed that the discount rate was $r = 0.05$. Using MATLAB to calculate optimal rotation times, we found that the optimal one-shot harvest length was $T^* = 10.69$, and that the optimal repeated rotation length was $T^{**} = 7.58$ years. As expected, considering multiple harvests shortens the optimal rotation period. Figure 3.4 shows this example, replicating the qualitative results of Fig. 3.3.

Although here there are two solutions to the first order conditions for an optimum in the repeated harvest problem with logistic growth, the one on the right is the global optimum, as shown by the level of the value function. Another way of understanding this result is that the first intersection cannot be optimal as beyond this point the asset is growing at a rate greater than the alternative rate of return, so the trees should be left in the ground—until the second intersection.

[2]We obtained these parameter values by fitting a logistic growth function to the data in Sect. 4.4 in Conrad (2010) on Douglas fir grown in the Pacific Northwest.

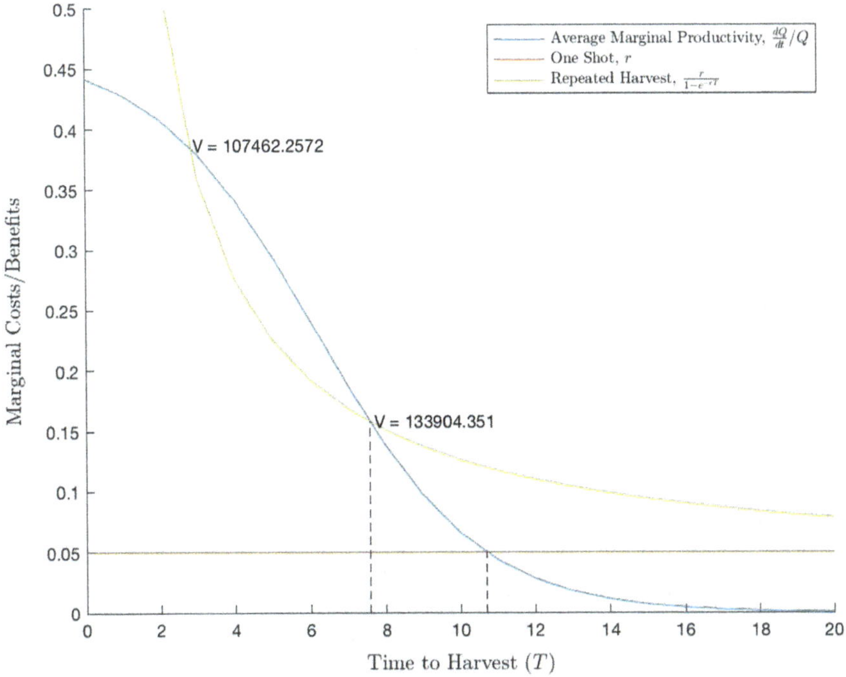


**Fig. 3.4** The numerical analog of Fig. 3.3, showing the difference in the optimal harvest dates

The optimal stopping times in our example are quite different to the times obtained in Conrad's original example. His analysis found an optimal one-shot harvest length of $T^* = 62.62$, and an optimal repeated harvest rotation length of $T^{**} = 61.12$ years. The difference arises because he assumes that the trees grow according to an exponential growth law (where total biomass is given by $Q(t) = e^{a-\frac{b}{t}}$). Although there is a large difference between the optimal harvest times, the qualitative result that the repeated harvest rotation length is shorter than the one-shot rotation length still holds. The two examples do however illustrate the sensitivity of the quantitative results of bioeconomic models to assumptions about the biological growth function.

### *3.2.4 Biological "Optimum": The Maximum Sustainable Yield (MSY)*

The problem can be stated as: Find the rotation period $T$ that maximizes average annual yield
Solution:

$$\max_T V = \frac{Q(T)}{T}$$

$$\frac{d\left[\frac{Q(T)}{T}\right]}{dT} = \frac{-Q(T)}{T^2} + \frac{1}{T}\frac{dQ}{dT} = \frac{-Q(T) + T\frac{dQ}{dT}}{T^2} = 0$$

$$Q(T) = T\frac{dQ}{dT}$$

$$\frac{dQ}{dT}\bigg/ Q = \frac{1}{T}$$

To compare to bio-economic optima, let

$$f(r) = r, g(r) = 1 - e^{-rT}$$

$$\lim_{r\to 0}\frac{f(r)}{g(r)} = \frac{0}{1-e^0} = \frac{0}{1-1} = \frac{0}{0}$$

By L'Hopital's Rule,

$$\lim_{r\to 0}\frac{f(r)}{g(r)} = \lim_{r\to 0}\frac{f'(r)}{g'(r)} = \lim_{r\to 0}\frac{\frac{dr}{dr}}{\frac{d(1-e^{-rT})}{dr}} = \lim_{r\to 0}\frac{1}{Te^{-rT}} = \frac{1}{T}$$

$$\text{In the limit, as } r \to 0,\ \frac{r}{1-e^{-rT}} \to \frac{1}{T}$$

This is a perhaps surprisingly simple result, given the somewhat involved manipulation needed to obtain it. It is also an intuitive result: with no discounting, an economic concept, the bio-economic optimum converges to the purely biological, the MSY, as illustrated in Fig. 3.5.

For the numerical example in Sect. 3.2.3, the MSY harvest length is 8.47 years. Figure 3.6 shows the optimal harvest time for the one-shot and repeated harvest problems as a function of the discount rate. As the discount rate approaches zero the optimal repeated harvest length approaches $\tilde{T} = 8.47$—illustrating the result we just derived. The one-shot harvest time approaches infinity as the discount rate approaches zero. This is because the growth rate for a positive stock of trees is only zero at the maximum stock size, which is only approached as $t \to \infty$. Put another way, a discount rate of zero means that the manager is completely indifferent between profit now and profit later, and if the stock is always growing, the present value of total profit can always be increased by waiting before harvesting.

As the discount rate increases, the return to non-forestry investment increases. For the logistic growth function there is a threshold discount rate above which investing in forestry is no longer optimal, i.e. trees never grow fast enough to compensate the forester for the opportunity cost of his capital. For the one-shot problem, an increase in the discount rate increases the height of the orange 'one-shot' curve in Fig. 3.4. The threshold discount rate for the one-shot problem in this example is

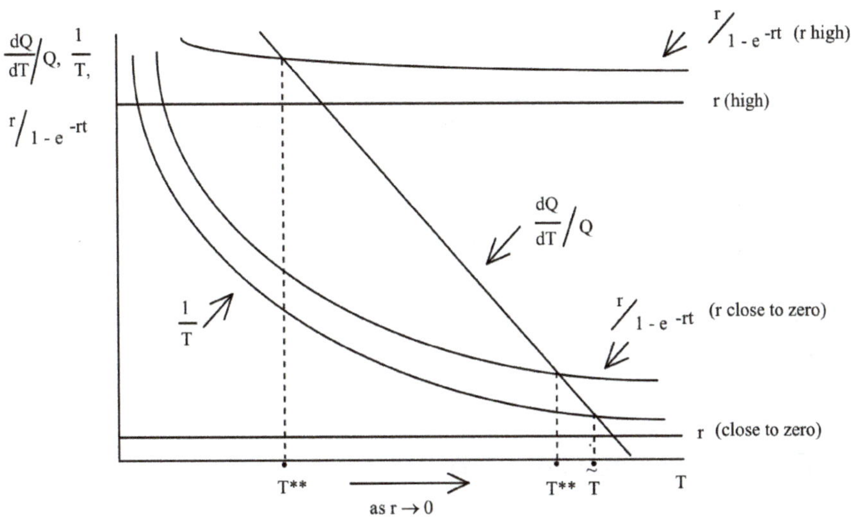


**Fig. 3.5** With no discounting, the bioeconomic optimum ($T^{**}$) converges to the purely biological maximum ($\tilde{T}$)

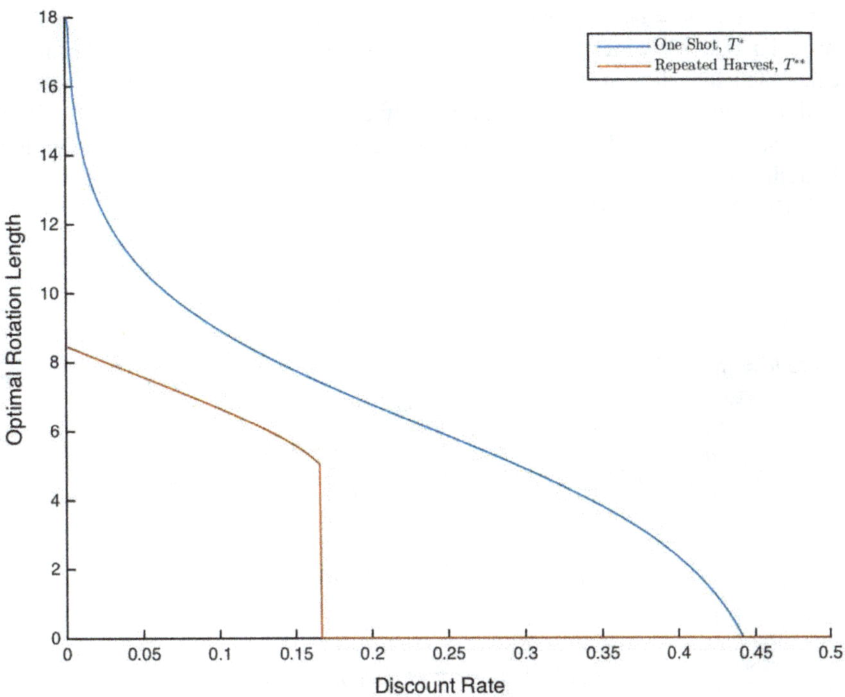


**Fig. 3.6** Optimal rotation length for one shot and repeated harvesting

0.43, and the optimal rotation length moves smoothly to zero as the discount rate increases. For the repeated harvest problem an increase in the discount rate shifts the yellow 'repeated harvest' curve from Fig. 3.4 to the right. As the discount rate increases eventually there is a point of tangency between the repeated harvest curve and the average marginal productivity curve, which occurs at the threshold discount rate. In this example the threshold discount rate for the repeated problem is 0.17. Figure 3.6 shows the optimal rotation age as a function of the discount rate. As the discount rate approaches zero, the repeated harvest solution converges to the Maximum Sustainable Yield rotation length of 8.47 years, while the one-shot length approaches infinity. Note the discontinuity in the repeated harvest optimal rotation age curve as the discount rate increases.

### 3.2.5 The Standing Forest Has Value

For simplicity, we return to the case of the one-shot harvest, with price unchanging over time. The results we shall obtain go through for the case of repeated harvests (Hartman 1976). We let $b_0$ be the initial value of the standing forest, value growing at rate $g$, i.e. $b_t = b_0 e^{gt}$. These assumptions can of course be modified to suit a particular case for which they are not appropriate. We make them here because it seems plausible that an old-growth forest, for example the redwoods of northern California, yields greater value than a newly planted, or younger forest. It will be obvious, as we proceed, how results can be affected by modifying the assumptions about the initial value of a standing forest and how this value evolves over time.

The problem can be stated as follows:

$$\max_T V = pQ(T)e^{-rT} + \int_0^T b_0 e^{ht} dt$$

where $h = g - r$

$$\begin{aligned}
\frac{dV}{dT} &= pQ(T)\left(-re^{-rT}\right) + e^{-rT} p\frac{dQ}{dT} + \frac{d}{dT}\int_0^T b_0 e^{ht} dt \\
&= -rpQe^{-rT} + p\frac{dQ}{dT}e^{-rT} + \frac{d}{dT}b_0 \left.\frac{e^{ht}}{h}\right|_0^T \\
&= -rpQe^{-rT} + p\frac{dQ}{dT}e^{-rT} + b_0\frac{d}{dT}\frac{e^{hT}}{h} - \frac{e^0}{h} \\
&= -rpQe^{-rT} + p\frac{dQ}{dT}e^{-rT} + b_0 e^{hT} = 0
\end{aligned}$$

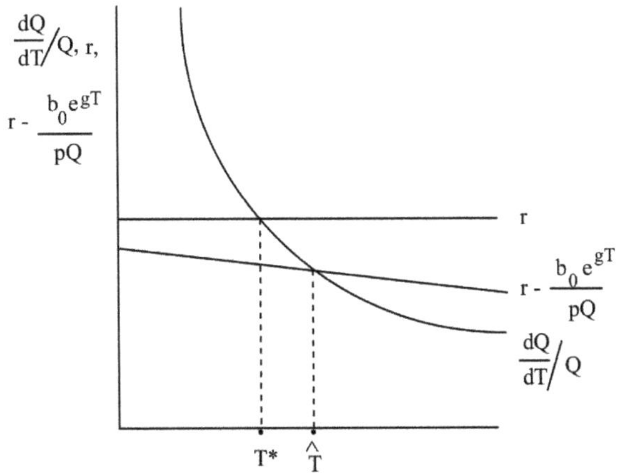


**Fig. 3.7** The optimal harvest date ($\hat{T}$) when the standing forest has value

$$p\frac{dQ}{dT}e^{-rT} = rpQe^{-rT} - b_0e^{hT}$$

$$p\frac{dQ}{dT} = rpQ - b_0e^{gT}$$

$$\frac{p\frac{dQ}{dT}}{pQ} = r - \frac{b_0e^{gT}}{pQ}$$

$$\frac{dQ}{dT}\Big/Q = r - \frac{b_0e^{gT}}{pQ}$$

$$\text{We know that } \frac{b_0e^{gT}}{pQ} > 0, \text{ so } r - \frac{b_0e^{gT}}{pQ} < r$$

Further, as shown in Fig. 3.7, the curve $r - \frac{b_0e^{gT}}{pQ}$ slopes downward since $b_0e^{gt}$ grows exponentially while $pQ$ grows more slowly over time. Note that, if the initial value of the standing forest is high enough, or is growing rapidly enough, or a combination of both, harvesting the trees may never be optimal. This may be the case for example for the old-growth redwood forest.

## 3.3 Optimal Control of a Generic Renewable Resource*

Consider the optimal management of a generic renewable resource. Using the notation from the optimal control model for an exhaustible resource (Sect. 2.5), with the addition of a biological growth function, $g(x_t)$, the problem can be written as

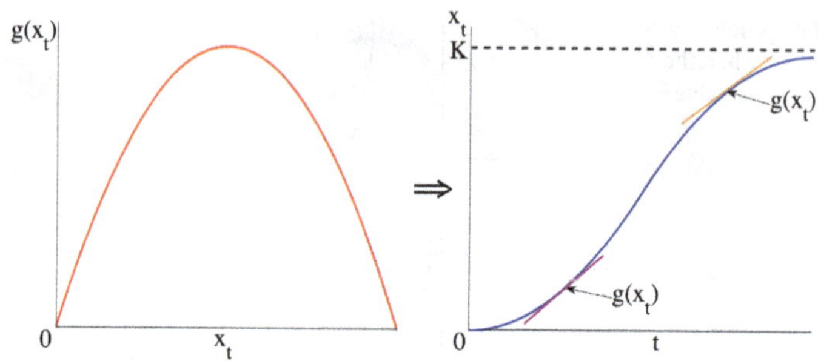


**Fig. 3.8** The growth function

$$\max \int_0^{\infty} [p_t q_t - c(q_t, x_t)] e^{-rt} \mathrm{d}t$$

$$\text{s.t.} \quad \dot{x} = g(x_t) - q_t$$

$$\text{where} \quad g(x_t) = \gamma x_t \left(1 - \frac{x_t}{K}\right)$$

Here $g(x_t)$ follows the logistic growth law, as used for example in the forestry problem, to characterize the behavior of a renewable natural resource or population where $\gamma$ is the intrinsic growth rate of the resource and $K$ is the carrying capacity of the environment. As noted earlier, for $x_t$ very small, $g(x_t) \cong \gamma x_t$; exponential growth. As $x_t \to K, g(x_t) \to 0$. These relationships are illustrated in Fig. 3.8.

The optimal control problem can be expressed as:

$$\tilde{H} = p_t q_t - c(q_t, x_t) + \mu_t (g(x_t) - q_t)$$

$$\frac{\partial \tilde{H}}{\partial q_t} = p_t - \frac{\partial c}{\partial q_t} - \mu_t = 0$$

$$\dot{\mu} = r\mu_t - \frac{\partial \tilde{H}}{\partial x_t}$$

$$= r\mu_t + \frac{\partial c}{\partial x_t} - \mu_t g'(x_t)$$

$$\frac{\dot{\mu}}{\mu} = r + \frac{\partial c}{\partial x_t} \Big/ \mu - g'(x_t)$$

This result, as in the exhaustible resource problem, has an economic interpretation. The new term on the r.h.s., $g'(x)$, provides an extra dividend, the extra growth that results from holding an extra unit of the resource in the stock.

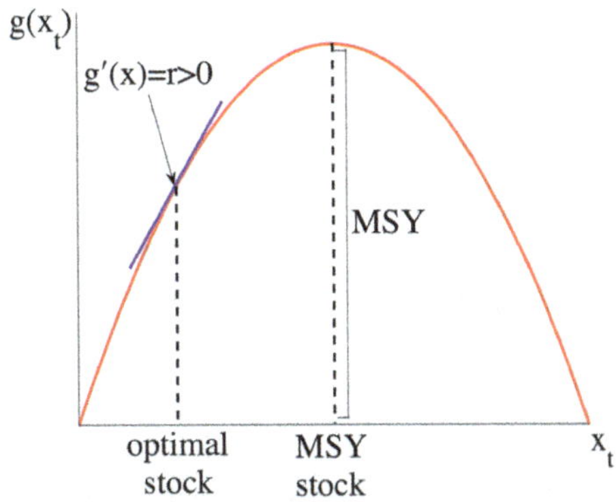


**Fig. 3.9** When the discount rate is positive, the optimal stock size is less than the MSY stock size

In the steady state, $\dot{\mu} = 0$, so

$$r = g'(x_t) - \frac{\partial c}{\partial x_t} \bigg/ \mu$$

$$\text{If} \quad \frac{\partial c}{\partial x_t} = 0, \; r = g'(x_t)$$

This is the traditional result, but note that it depends on the absence of a stock effect on extraction or harvest cost (much like the "*r* percent rule" for nonrenewable resources does).

Assuming $r > 0$, this also implies that the steady state stock corresponds to $g(x)$ increasing ($g'(x) > 0$), or in other words the stock is to the left of (is less than) the maximum sustainable yield (MSY) stock, illustrated in Fig. 3.9.

Of course, if $r = 0$, then $g'(x) = 0$, and the MSY stock is optimal, as in Fig. 3.10. We can characterize the MSY stock as follows:

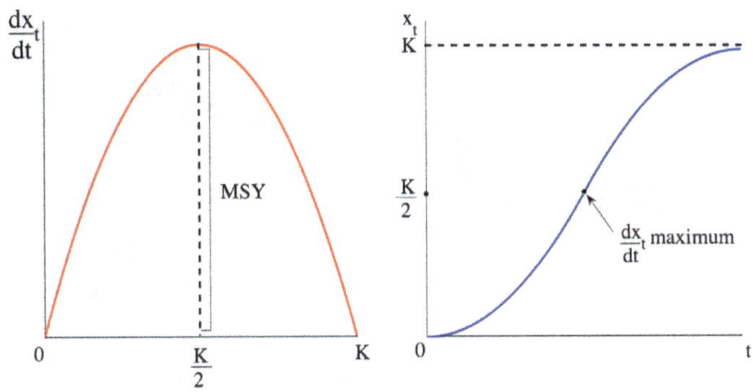


**Fig. 3.10** The MSY stock is optimal when the discount rate is zero

$$\begin{aligned} g(x_t) &= \frac{\mathrm{d}x_t}{\mathrm{d}t} = \gamma x_t \left(1 - \frac{x_t}{K}\right) \\ &= \gamma x_t - \frac{\gamma}{K} x_t^2 \\ \frac{\mathrm{d}\left(\frac{\mathrm{d}x_t}{\mathrm{d}t}\right)}{\mathrm{d}x_t} &= \gamma - \frac{2\gamma}{K} x_t = 0 \\ \frac{2\gamma}{K} x_t &= \gamma \\ x_t &= \frac{K}{2} \end{aligned}$$

The result that the MSY stock is optimal has a nice, intuitive interpretation: In the absence of discounting the bio-economic optimum is equivalent to what might be called the purely biological optimum, the MSY. Recall that we obtained essentially the same result in the timber harvesting problem.

# Chapter 4
# Environmental Resources: Dynamics, Irreversibility and Option Value

**Abstract** This chapter begins with a short discussion of the evolution of environmental economics, with links to theories of externalities and public goods and a focus on pollution. The focus then shifts to a less-studied aspect: the economics of natural environments. The idea that the environment from which commercially valuable resources are taken may have value in its natural state motivates a discussion of how this affects a benefit/cost analysis of the resource development project. This in turn leads to the concept of irreversibility, joined to uncertainty concerning evaluation of the environmental services that would be lost in perpetuity, or at least in the very long run if the resource development project goes forward. Examples are given to motivate the realism and potential importance of this approach. A theoretical discussion of investment under uncertainty and irreversibility is presented, verbally and graphically, which leads to the concept of option value (real, not financial), in turn given an analytical treatment, including an example of how to compute. The potential empirical importance of option value is illustrated in an application to a forested area in Thailand, in which each of four zones can be either preserved, developed, or given an intermediate level of development, in each of three periods. The states are distinguished by appropriate uses but subject to an irreversibility constraint, formulated as a decision tree of feasible sequences. For example, commercial agriculture can follow preservation but not vice versa. Results of an exercise using illustrative values show that option value, though small in relation to total use values, can tip the balance in a major way, with the largest zone optimally developed in what might be called the traditional analysis but preserved when option value is taken into account.

## 4.1 Introduction: The Transition from Extractive to in Situ Resources

In our discussion of the roots of resource economics we noted, and subsequently formalized in developing the theory of optimal depletion, two ideas from Mill (1848) that have a contemporary ring: (1) Extraction costs will increase as mineral deposits are depleted, owing to the need to sink shafts deeper, work thinner seams, move to


A. C. Fisher, *Lecture Notes on Resource and Environmental Economics*, The Economics of Non-Market Goods and Resources 16,
https://doi.org/10.1007/978-3-030-48958-8_4

lower quality deposits, and so on. (2) The increase will be mitigated by discovery and technical change. Mill had another modern idea: that a stock of land was valuable not only for what could be extracted from it but also for the opportunities it provided for experiencing natural beauty and solitude.

Whether this had any influence on resource policy is not clear, but Mill's treatise had been through several editions by the time the national park system was established in the United States, in 1872, and preservation of natural beauty in wilderness environments was a minor goal of the early conservation movement of 1890–1920. But the theme was not picked up by the economics profession until very much later. Instead, environmental disruption was first analyzed as a static externality, following the work of Pigou (1932) and his examples of sparks from railway engines, factory smoke, and the like. By the 1950s, several economists were developing elements of the modern theory of externalities, but in the process losing sight of Pigou's environmental examples, rather focusing on external economies in production, occasionally external diseconomies in production. Largely ignored was the direct interaction between one or many producers, on the one hand, and large numbers of consumers, on the other, that characterizes air and water pollution and other forms of environmental disruption we study today.

Not until the early 1960s was the Pigouvian approach applied in a systematic fashion to the problem of dealing with water pollution, and later air pollution. Credit for this achievement probably is due to economists in the water resources program at Resources for the Future (RFF), notably Allen Kneese (1964). More recently, Mill's somewhat different concern for the preservation of natural environments has been embodied in several studies extending the traditional benefit-cost analysis of water-resource and other development projects. This line of research stems from what I would characterize as the other great contribution of the early years of RFF, John Krutilla's (1967) reconsideration of the traditional concerns of conservation, which established the conceptual framework for the subsequent benefit/cost analyses.

This chapter will deal with the preservation of natural environments or, not to prejudge the issue, with the allocation of in situ environmental resources rather than the more traditional topic of pollution externalities. Suppose that the proposed site for a planned development project, such as a dam for hydroelectric power or an open-pit mine for molybdenum (to take two examples that have received attention in the literature), can also yield value in its natural state. How should this affect the benefit-cost analysis of the project? Further, what special problems are posed by the prospect that the in situ resource cannot be reproduced once it has been lost or destroyed in the process of development? The strong concern felt by many for the fate of threatened environments and their indigenous species presumably reflects a perception that loss will be irreversible. Assuming that this is true, what changes in project investment rules are called for? Questions can also be asked about the implications of uncertainty. Knowledge of the values to be attached to long-lived extra-market consequences is bound to become unreliable as one looks further into the future. Again, how can this be dealt with analytically? In this chapter we consider these and related questions.

## 4.2 Irreversibility in Economics and in Environmental Processes

Because the assumption that conversion of a natural environment is irreversible drives the results we shall discuss, let us explore it in some detail. One might take the view that any investment is irreversible, in the sense that time does not move backward. On the other hand, an economist's view might be that the consequences of any decision, including a decision to develop a natural environment, are reversible given sufficient input of labor and capital. I believe neither view is helpful in determining how to use the resources of a natural environment. Meaningful distinctions can be made between uses that are reversible and those that, for all practical purposes, are not. The discussion here draws heavily and very directly on Krutilla's seminal work and also on a comprehensive volume on the work of the Natural Environments Program at RFF (Krutilla and Fisher 1975) and a volume covering more broadly topics in resource and environmental economics (Fisher 1981).

Consider, for example, the use of resources that represent an accident of geologic processes—the geysers in Yellowstone National Park. These can serve either as a source of geothermal energy for the production of electricity or as a basis for tourism and related recreational and perhaps scientific activities. It is true that Yellowstone National Park represents a serious commitment to preservation of the natural features, but should the continued growth of the U.S. economy depend on its being reoriented to another use, no technical constraint would prevent this. If these geothermal resources should be allocated to energy production, on the other hand, the consequences to the environment would be virtually impossible to reverse. Construction of plants to generate electric power by use of the geothermal steam, along with the associated switchyards, transmission towers, and so on, would result in an adverse modification of the scenic environment in the park for a very long time, if not permanently. The mining of the superheated water would, in sufficient time, reduce subsurface pressures, eliminate the geyser phenomenon, and thus remove the reason for the establishment of the area as a national park. Any attempt to "restore" the area following depletion of the geothermal resources would be technically impossible.

More traditional mining projects also pose problems, especially in high mountain or arctic environments. Removal of the primitive vegetal cover to expose mineral earth can lead to increased absorption of solar heat, which in turn affects unstable soil relationships in areas of permafrost, with resultant thawing, erosion, and gullying—and as recent discussion of climate change indicates, potential release of the potent greenhouse gas methane in arctic tundra regions (Fisher and Le 2014b).

The biological environment can also be adversely and irreversibly affected. An obvious example is the loss of an entire species and the genetic information it contains, should its essential habitat be destroyed. Even if species survival is not at issue, biological impacts can be very difficult to reverse over any time span that is meaningful for human societies. The clear-cutting of a climax species is equivalent to removing the results of an ecological succession that may represent centuries of natural processes. The removed climax species may be succeeded by others in a procession of changing plant and animal communities, culminating in the original ecological relationships only after a very long time, well beyond the horizon of traditional economic or planning models.

But might it not be technically feasible, with sufficient input of conventional factors of production, to short-circuit these very slow natural processes in some cases? We have thus far considered cases in which this would not be possible. Others that involve relatively ordinary landscapes in humid or sub-humid zones may be candidates for some form of restoration. But in evaluating any such restoration one must also take account of the preferences of users. No matter how skillfully Disneyland simulates an environment, devoted Sierra Clubbers may not be satisfied. This is not intended as a criticism of Disneyland or of the Sierra Club. Authenticity in a natural environment is, to some, a valued attribute, just as authenticity in a work of art is to others. And in assessing the value of a resource or a painting there is no obvious reason to overlook the preferences of the "purists." The preferences of wilderness recreationists have been studied extensively and results suggest the existence of a substantial number of people who value highly the attribute of authenticity in an environment. Moreover, the prospect is that this number will grow, because such preferences appear to be positively correlated with income and education levels.

Suppose it is accepted that, to a first approximation, the economic development of a natural environment that is in some way remarkable will be irreversible. Are we really saying anything more than we have already said about exhaustible extractive resources? In my judgment we are. The in situ resources of the environment may be "more exhaustible" than conventional exhaustible resources. Consider an open-pit mine in a scenic area. The final consumer of the mine's output is presumably indifferent to the source. Although depletion of a particular deposit is irreversible, this may not matter much if other deposits can be made available. Moreover, because the mineral output will tend to enter production as an intermediate good, lower-grade deposits or even other minerals could easily substitute (think of aluminum for copper). In fact, as we shall discuss in a later chapter, technical change in the extractive and logistic-support industries made possible the production of mineral commodities at generally declining relative supply prices over a century of industrial growth in the U.S. and other developed countries. But technology can do little to reproduce the results of the particular patterns of geomorphology, weathering, and ecological succession found in the scenic environment in which the mineral deposit occurs. The amenity services of the environment tend to enter directly the utility functions of consumers, with no intervening production technology. When there are perceived differences between this environment and others, as in general there will be, perfect substitution (in consumption) is not possible, and loss of a particular environment may matter, at least to some.

## 4.3 Evaluating Irreversible Investments

To economists, the important question about irreversibility is: What are the implications for resource allocation? If the in situ resources of an environment are declining in value relative to the extractive resources, then clearly irreversibility poses no special problem. An optimal investment program will call for conversion at a rate dictated by the changing relative values. Unfortunately, just the reverse may well be true. Unique natural environments are in many cases likely to appreciate in value relative to goods and services they might yield if developed. Then the restriction on reversibility matters, because value would be increased by going back to an earlier, less developed state.

An argument about technical change, relative values, and irreversibility might go something like this. Technical change is asymmetric. It results in expanded capacity to produce ordinary goods and services, but not natural environments. As long as consumer preferences do not shift sufficiently in favor of the ordinary goods (and we have evidence that they are likely to shift in the opposite direction), the supply shift implies an increase in the relative value of the in situ resources (Smith 1974). This is pertinent to the assessment of any proposed conversion of the resources (the construction of a large dam, say, or an open-pit mine). Because the value of the in situ resources may be increasing relative to that of the water, power, or minerals produced by the development project, and development is irreversible, we might reasonably expect project investment criteria to be somewhat conservative.

One obvious way to proceed would be simply to extend the model of optimal depletion to take account of environmental costs. These could be related to the rate of depletion or the size of the stock or both. But recall that costs are already functions of both. It is easy to verify that adding an environmental component will increase (social) costs of depletion in each period, in turn resulting in a reduction in the optimal rate of depletion. But this result does not fully capture the effects of irreversibility, because these have more to do with the development of an area for extractive production than with the subsequent rate of extraction.

A model of irreversible development of a natural environment, based on Arrow's analysis (Arrow 1968) of a formally similar problem of optimal capital accumulation with irreversible investment, demonstrates that if there is a period of time over which it is anticipated that the benefits from development net of environmental costs will be declining, then development is optimally cut off some time before the start of that period, as represented in Fig. 4.1 (Fisher et al. 1972).

The intuition behind these results is clear. As long as development is irreversible, some account must be taken of the behavior of future costs and benefits. If it is anticipated that these may be changing in the way we have proposed, the optimal level of development is decreasing, over an interval of time that may stretch to the planning horizon. Then some near-term losses from having too little development are absorbed in order to avoid later losses from too much development. Further, in the case in which expected net benefits are always declining, either no development is appropriate or development is undertaken at the initial point in time, when conditions

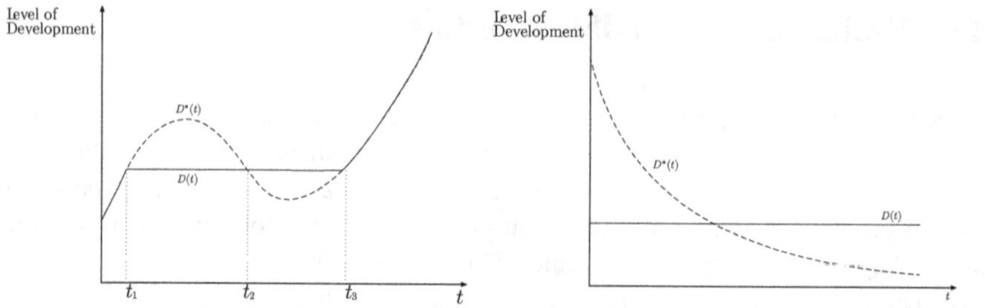


(a) Investment is halted between $t_1$ and $t_3$. Since investment is irreversible, the optimal path $D(t)$ underinvests during the interval $[t_1, t_2]$ to avoid excessive overinvesting during the interval $[t_2, t_3]$.

(b) If the myopic investment path is always declining, the irreversible investment path, $D(t)$, will be set a constant level over the planning horizon.

**Fig. 4.1** Development paths for an irreversible investment. Path $D^*(t)$ is the 'myopic' path, not taking irreversibility into account. Path $D(t)$ accounts for irreversibility

are most favorable. Note that the same project assessed at a later date may look better than the alternative of no project. But the point is that it would have been even better to have undertaken it sooner.

This conclusion could, of course, be upset by changes in the behaviors of the cost and benefit streams. In the model we have been discussing, this is not possible. Expectations about benefits from both of the alternative uses of the environment are assumed not to change with the passage of time. Yet, clearly, people do learn. Prior probability distributions are revised in the light of new information. This seems particularly relevant in the case of the long term behavior of the benefits of environmental preservation, a non-market good with highly uncertain—though potentially large—benefits even in the short term.

## 4.4 Investment Under Uncertainty and Irreversibility

Suppose you may wish to visit the Grand Canyon at some time in the future. Would you be willing to pay anything to ensure the future availability of the canyon, or in other words, to retain the option of visiting? There is, of course, your consumer surplus from the visit. But is the value of the option just expected consumer surplus? Suppose you are risk-averse. Then you would be willing to pay a premium in addition to the expected return. One concept of option value is that this premium is, in fact, option value—the difference between (a) what one is willing to pay for the option of consuming (at a predetermined nondiscriminatory price) in the future, and (b) expected consumer surplus. Weisbrod (1964) was probably the first to suggest that when the demand for a publicly provided good is uncertain, there may be value in retaining the option to consume, apart from conventional consumer's surplus. Following some controversy in the literature, and Cicchetti and Freeman (1971) demonstrated that option value can be identified with a risk premium over and above expected consumer's surplus.

There are, however, a couple of difficulties. In the first place, preservation can also bring risks: the risk of flood, for example, or of power failure. The net option value of preserving a wilderness environment could then be negative Henry (1974). Further, it is not clear that a social choice should display aversion to risk, even though the affected individuals do. As noted earlier in the discussion of private and social discount rates, two arguments have been advanced in support of this view. First, if the government undertakes a great many investment projects, risks may be pooled. Second, if the net returns to a single investment are spread over a great many individuals, so are the risks. In either case, the risk premium disappears.

Suppose we accept the view that a social choice about the economic development of a natural environment ought to be risk-neutral, or that, even if it is not, there are risks on both sides. Can we establish the existence of an option value of preserving the environment? It turns out that the asymmetry of alternatives (one is irreversible, the other is not) is crucial. We shall now assume that the passage of time results in new information about the benefits of each. But this new information can be taken into account only to the extent that development has not already occurred. Once it has, information that suggests it would be a mistake cannot affect the outcome. Accordingly, there is some value in refraining from an irreversible action that otherwise looks profitable. This, at any rate, is the interpretation of a result we shall now obtain.

Better information can improve decision-making. But investing in information has a cost: there are either direct costs in the form of expenditures involved in conducting a research project, or an opportunity cost to refraining from the development in question and waiting for information to accrue over time. To know whether to wait, as opposed to going ahead with an uncertain venture, we need to know the benefit of waiting: the value of information, defined as the difference between the expected value of the optimal decision with the new information and the expected value of the optimal decision without the information. To determine whether or not to invest in information, it is necessary to compare the value with the cost of acquiring the information, either the direct cost of research or the opportunity cost of waiting or both.

We can then define an alternative concept of option value as the value of information conditional on retaining the option to make an irreversible decision (or not) in the future (Hanemann 1989, based on earlier results of Arrow and Fisher 1974 and Henry 1974). The concept seems straightforward but the computation perhaps less so, as we illustrate with a numerical example followed by an illustrative empirical application.

## 4.5 Computation of Option Value

Let $S_1$ be the fraction of an area developed in the first period, $S_2$ be the fraction of an area developed in the second period, and $b_1$ be benefits in the first period net of preservation benefits. (For example, if the benefits of development are 100 and the benefits of preservation are 80, then the net benefits $b_1 = 20$.) Assume that $b_1$ is defined for $S_1 = 1$ so that, in general, first-period benefits are $S_1 b_1$. Finally, assume that $b_1$ is known at the start of the first period and that $b_2$ is a random variable with a known distribution. For example,

$$b_2 = 100 \text{ with probability } p = \frac{1}{2}$$

$$b_2 = -50 \text{ with probability } 1 - p = \frac{1}{2}$$

$$\text{So } E[b_2] = \frac{1}{2}100 + \frac{1}{2}(-50) = 25$$

Now, we set up two "information structures".

### 4.5.1 *No Expectations of Learning*

Assume that no further information about $b_2$ will be available at the start of the second period when we need to choose $S_1$. (Alternatively, assume that we simply do not take account of the prospect of better information.) We then choose $S_1$ and $S_2$ at the start of the first period with the choice of $S_2$ based on the expected value of $b_2$.

Since $E[b_2] > 0$, we chose $S_2 = 1 - S_1$, i.e. fully develop in the second period. We make the decision on the basis of *expected value* in the second period, i.e., we choose the $S_2$ that maximizes the expected value $E[b_2](S_1 + S_2)$, in this case $25(S_1 + (1 - S_1))$.

Now, consider the choice of $S_1$. Maximize benefits over the first period (and, indeed, over both periods, given the choice of $S_2 = 1 - S_1$ in the second period) by choosing $S_1$ to maximize:

$$V^* = S_1 b_1 + E[b_2](S_1 + S_2) = S_1 b_1 + 25(S_1 + (1 - S_1)) = S_1 b_1 + 25$$

The decision rule is then: $S_1 = 0$ if $b_1 < 0$, and $S_1 = 1$ if $b_1 > 0$.

Note that the "bang-bang" decision rule follows from linearity of the benefit function.

### 4.5.2 Anticipating Learning

Now assume that better information about $b_2$ will be available by the start of the second period. In particular (though this is not necessary for the result that we shall obtain), assume that $b_2$ will be known. We can take account of the new information by choosing $S_2$ at the start of the *second* period. Thus, we no longer simply substitute the expected value, $E[b_2]$, and make a decision on $S_2$ on the basis of this expected value at the start of the first period. Suppose that $b_2 = 100$. Then we choose $S_2 = 1 - S_1$, and benefits in the second period are $[S_1 + (1 - S_1)]100 = 1 \times 100 = 100$.

Now, suppose that $b_2 = -50$. Then we choose $S_2 = 0$, and benefits in the second period are $(S_1 + 0)(-50) = -50S_1$. Remember, development is irreversible so we are left with the $S_1$ chosen in the first period even though it is now unwanted.

How do we (optimally) choose $S_1$ under these circumstances, i.e., to maximize value over both periods? Note that we are, again, maximizing expected value since there is a link to the uncertain second-period benefits but, now, the *expected value of the maximum* rather than the *maximum of the expected values*. Thus we wish to maximize:

$$\begin{aligned}\hat{V} &= S_1 b_1 + E[\max\{b_2(S_1 + S_2)\}] \\ &= S_1 b_1 + \frac{1}{2}(100) + \frac{1}{2}(-50S_1) = S_1 b_1 + 50 - 25S_1 \\ &= S_1(b_1 - 25) + 50\end{aligned}$$

At the start of the first period, we have an expected value of what second-period benefits will be, given the optimal choice of $S_2$ on the basis of the information available at the start of the second period.

The decision rule is then: choose $S_1 = 0$ if $(b_1 - 25) < 0$, i.e. if $b_1 < 25$, and $S_1 = 1$ if $(b_1 - 25) > 0$, i.e. if $b_1 > 25$.

This is a more conservative decision rule than in the no-information case. We only develop the area if the first-period benefits are greater than some positive number (in this example, 25) rather than greater than zero. Since development is irreversible, once a decision to develop has been made, it cannot be affected by new information that suggests that it was a mistake. So there is a benefit to waiting, to refraining from the irreversible alternative.

Note that the benefit to following the second decision rule, based on taking account of the prospect of new information, must be as least as great as the benefit to following the first. The difference is the *value of information*, $\hat{V} - V^*$ in our example, where:

$$\hat{V} - V^* = (S_1 b_1 + 50 - 25S_1) - (S_1 b_1 + 25) = (50 - 25S_1) - 25 = 25(1 - S_1) \geq 0$$

If $S_1 = 0$, then $25(1 - S_1) = 25$. This is the *option value*: the *value of information conditional on retaining the option to invest (or not) in the future*. On the other hand,

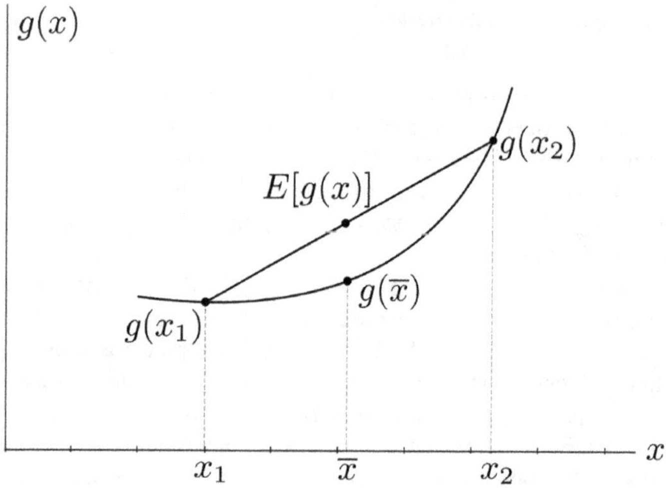


**Fig. 4.2** The expectation of a convex function is strictly greater than the function of the expectation

if $S_1 = 1$, then $25(1 - S_1) = 0$. Once the investment is made, there is no benefit from information that will become available in the second period.

This result is of course specific to the numerical example. But it can be shown that it holds more generally. Let $g(x)$ be a convex function of the random variable $x$. By Jensen's Inequality, $E[g(x)] \geq g(E[x])$, i.e. the expected value $E$ of a strictly convex function of a random variable is strictly greater than the convex function of the expected value, $E[x]$.

This sounds complicated but is easily illustrated with a diagram, as in Fig. 4.2. To simplify the illustration with no loss in generality, suppose the random variable can take just two values, $x_1$ and $x_2$. Writing $\overline{x}$ for $E[x]$ to avoid clutter on the figure, we see that, for any value of $x$ between $x_1$ and $x_2$ such as $\overline{x}$, $E[g(x)]$, a point on the line joining $g(x_1)$ and $g(x_2)$ lies above $g(\overline{x})$, a point on the curve representing the convex function $g(x)$.

What is the relevance of this relationship to our option value example? Although it is not usually thought of this way, the maximum function is convex. Thus letting $g(x) = \max(x)$, we can write

$$E[\max(x)] \geq \max E[x]$$

which is exactly what we have in $\hat{V} - V^*$ above.

## 4.6 Application of Uncertainty and Irreversibility: Valuation and Management of Tropical Forests

### 4.6.1 A Framework for Valuation

Tropical forests are being converted to one or another sort of commercial use in part because the value of a preserved forest is not getting factored into decisions on use. One obvious reason for this is that the value may not be captured by those who make decisions on use. Another reason is that the value of preservation, which is a non-market good, is hard to measure. This section, based on a study by Albers et al. (1996), sets up a framework for the valuation of tropical forests, draws on the concept of option value to deal with relevant uncertainty and irreversibility, and describes an application to a forest area in Thailand. We begin by considering the alternative uses of a tropical forest and the values they give rise to.

1. Uses compatible with preservation
   Hunting and fishing, gathering of food and forest products such as oils and medicines, and ecotourism are examples. Note that all are sustainable: the capacity to enjoy them is not diminished over time, due to the low-intensity, short-duration nature of the resulting disruption of the ecosystem.[1] Further, standing forests provide off-site environmental services, both local (protection against erosion) and global (carbon sequestration, conservation of genetic information), though the contributions of any one tract of forest land to the global public goods may be small, and in the empirical application we don't impute any value for this.
2. Intermediate uses
   Intermediate uses maintain some of the benefits of preserving forests while providing other economic benefits. Examples include agroforestry (crops planted among trees), selective harvesting of trees, and small-scale shifting cultivation. All can be sustainable.
3. Commercial forestry
   Commercial forestry is usually the first step in conversion of a tropical forest, but likely unsustainable. The soil is poor, with most of the nutrients stored in the vegetation and lost when trees are cut. Further, invasion of hardy grasses can out-compete early successional tree species, and even during the period of timber cropping the monoculture provides little of the original forest's ecosystem functions or related benefits.
4. Commercial agriculture
   Commercial agriculture includes plantation and livestock, as well as intensive shifting cultivation, characterized by insufficient time for land recovery and no buffer zones. Like commercial forestry, commercial agriculture may not be sus-

[1] Sustainable use is defined as use that maintains aggregate capital stock, including natural capital, implying an undiminished capacity to produce income.

tainable due to soil erosion, loss of nutrients, and buildup of fertilizer and pesticides that interfere with local environmental services.

5. Other large scale extractive activities
   Prominent examples are mining, which is by definition not sustainable, and water resource development, which can become unsustainable as reservoirs silt up. Both can however produce very high values in the near to medium term, and consequently be worth undertaking, depending on the comparative costs and benefits in a particular case.

Now we can relate these alternative uses of the forest to each other in a framework for valuation. We start by making a distinction between valuing specific services of the forest, and the forest itself, which can be viewed as an asset generating a stream of services over time. More precisely, the forest can be viewed as a portfolio of assets, whose composition can be varied over time, subject to some constraints. The constraints involve the sequencing of uses. Roughly speaking, conversion from a lower numbered use to a higher numbered use is feasible, but not vice versa. For example, livestock ranching may follow the clearing of land for a timber harvest, but not vice versa. Or commercial forestry may follow indigenous gathering, but not vice versa.

We can represent the alternative patterns of feasible sequences as in Fig. 4.3, collapsing the alternative uses to just three: *P* (preservation), *M* (intermediate), and

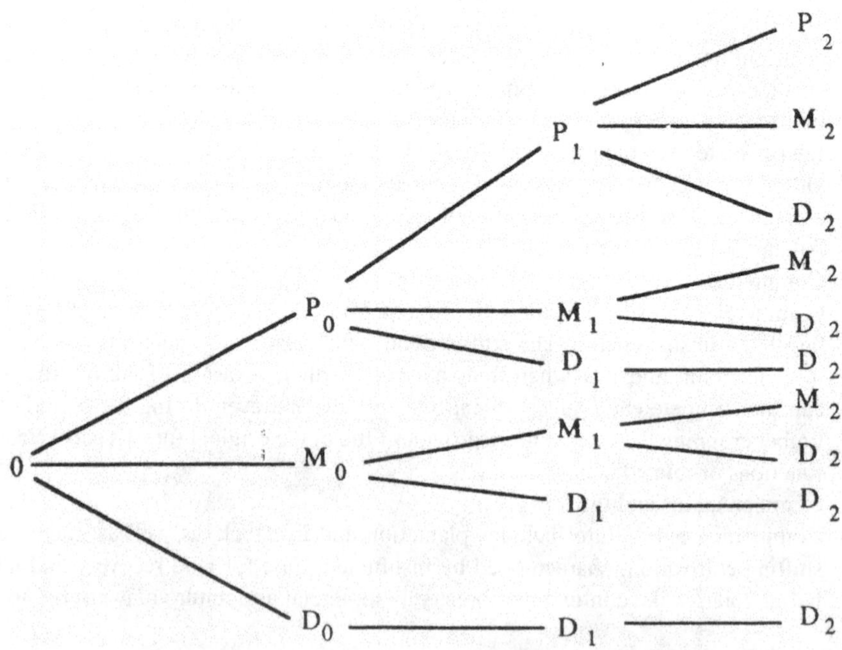


**Fig. 4.3** Feasible development sequences over three periods and three classes of use

$D$ (development). Thus $P$ could represent indigenous gathering, $M$ agroforestry, and $D$ commercial forestry. Note that $M$ and $D$ can follow $P$ and $D$ can follow $M$ but $P$ cannot follow $M$ or $D$ and $M$ cannot follow $D$. This represents an extension of the simple case described earlier in our example, in which there are two uses or goods, $P$ and $D$, in two periods. In the application we look at three periods and not just two to allow for different degrees of sustainability in activities such as timber harvesting.

How do we value the sequence of uses? Note first that the benefits of each use in the future periods will, in general, be uncertain. The usual way of dealing with uncertainty in a benefit/cost analysis is to replace the uncertain, or random, variable, say the benefits of preservation in the third period, with its expected value $E[P_2]$ where $E[\cdot]$ represents the expected present discounted value of the term in brackets. To avoid clutter in notation we are folding the discounting into the expression for expected value.

Suppose we look at the returns to a choice of preservation in the first period. Assuming the forest portfolio is managed to provide maximum benefits in the second and third periods, subject to feasibility constraints, the expected value of first-period preservation is

$$\begin{aligned} V_p^* = P_0 + \max\{&E[P_1] + \max\{E[P_2], E[M_2], E[D_2]\}, E[M_1] \\ &+ \max\{E[M_2], E[D_2]\}, E[D_1] + E[D_2]\} \end{aligned}$$

where $P_0$ is the known first-period benefit, $E[P_1]$ is the expected second-period benefit of preservation, and so on.

But now suppose we recognize that *information* about the returns to the different activities will be forthcoming, in the course of time and perhaps as a result of market or other research. The prospect of future information raises the possibility that different and presumably better decisions could be made about future uses when the time comes—if those uses have not been foreclosed.

Suppose, in particular, that at the start of each period we know the return to each of the alternative uses in that period (though not in the future). Then the expected present value of preservation in the first period is

$$\hat{V}_p = P_0 + \max\{P_1 + \max\{P_2, M_2, D_2\}, M_1 + \max\{M_2, D_2\}, D_1 + D_2\}$$

As in the numerical example, by Jensen's Inequality we have

$$\hat{V}_p - V_p^* \geq 0$$

or, in other words, the value of preservation in the first period is potentially larger given flexibility to use new information in making future choices.

The difference between $\hat{V}_p$ and $V_p^*$ is the value of information, in this case also the option value, the value of information conditional on retaining the option to decide on conversion to $M$ or $D$ in a future period when new information about returns will be available.

### 4.6.2 An Illustrative Application

The application, taken from Albers et al. (1996), is to a forest area in Thailand, part of which is currently designated as a park, Khao Yai National Park (KYNP). As the title of this section indicates, the application is intended only to illustrate the potential importance of the concepts we have been developing in this chapter. The problem of how to choose among competing uses of the area over time can be structured to focus on three uses ($P$, $M$, and $D$), over three periods, in four zones (Zones 1–3 within KYNP, Zone 4 proposed for KYNP). Each period is 12 years, the investment cycle for a eucalyptus plantation, the development use. There is uncertainty about returns to $P$, but not $M$ or $D$. This formulation is the simplest that will allow us to deal with issues of uncertainty, learning, and sustainability (Fig. 4.4).

The benefits from each use in each period are shown in Table 4.1.

Preservation benefits:

1. Erosion control: "cost of replacement" assessment of nutrient loss on nearby agricultural land from erosion if KYNP is not preserved
2. Hydrologic benefit: marginal return to irrigated land of controlling water flow to surrounding agricultural land
3. Tourism & recreation: consumer surplus for KYNP, based on data from similar areas
4. Extractive goods gathered in area: assumed sustainable; all values calculated for entire area and imputed for each zone based on existing uses, size, and geography.

Two states of the world define uncertainty about these values. The high state reflects additional value from a viable elephant population, growing at rate $r = 0.025$/year. The low state reflects the possibility of a drop in tourism in later periods. The probability of the high state is $p = 0.5$ in the second period and $p = 0.25$ in the third. The probability of elephant survival declines over time, and a threatened spread of AIDS could dampen tourism (perhaps less serious now; remember, this is an illustrative example).

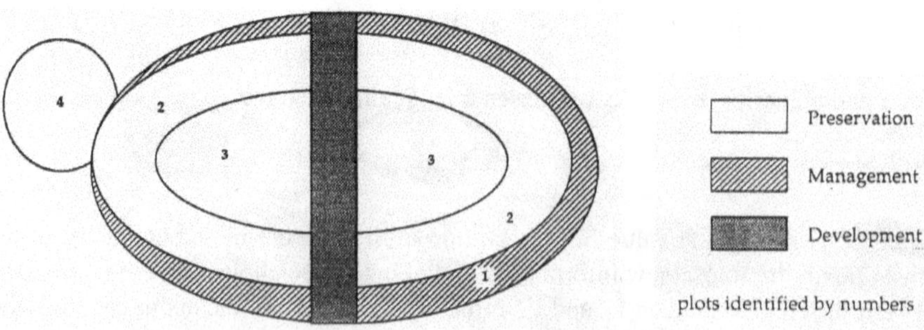


**Fig. 4.4** Khao Yai National Park (KYNP) plots and initial uses

**Table 4.1** Data for benchmark Khao Yai National Park

| Period | Use | Zone 1 | Zone 2 | Zone 3 | Zone 4 |
|---|---|---|---|---|---|
| Time 1 | $P$ | 29.04 | 777.42 | 653.54 | 14.35 |
| Time 2 | $P$ (low event) | 8.33 | 228.11 | 154.59 | 4.36 |
| | $P$ (high event) | 8.33 | 278.85 | 290.11 | 4.36 |
| Time 3 | $P$ (low event) | 2.64 | 72.37 | 49.05 | 1.38 |
| | $P$ (high event) | 2.64 | 94.63 | 108.49 | 1.38 |
| Time 1 | $M_a$ | 41.67 | 439.75 | 416.32 | 33.8 |
| Time 2 | $M_a$ | 13.28 | 140.12 | 132.65 | 10.77 |
| Time 3 | $M_a$ | 4.23 | 44.65 | 42.27 | 3.43 |
| Time 2 | $M_b$ | 8.16 | 86.15 | 74.92 | 1.13 |
| Time 3 | $M_b$ | 2.6 | 27.45 | 23.87 | 0.36 |
| Time 3 | $M_c$ | 1.46 | 15.36 | 10.96 | 0.1 |
| Time 1 | $D$ | 50.91 | 840.31 | 828.44 | 9.36 |
| Time 2 | $D$ | 16.41 | 269.07 | 265.34 | 2.98 |
| Time 3 | $D$ | 5.42 | 86.73 | 84.86 | 0.95 |

Note: All values are present discounted values (interest rate of 10%) of benefits received in the time period indicated in column 1. The uses $M_a$, $M_b$, and $M_c$ represent the intermediate use values as they decline over the amount of time in that use (one period, two periods, and three periods, respectively)

Intermediate use benefits:

1. Shifting cultivation: convert 10% of a zone's land to small scale agriculture, farm for two years, then abandon so at the end of a 12-year decision period, 60% of the land has been converted and is in various stages of recovery. Values drop off over time as more land is converted from forest, leaving less erosion control and fewer regenerative pathways
2. Extractive goods growth: more intense than in $P$, therefore not sustainable; values drop off over time as the area is being "mined".

Development benefits:

1. Permanent agriculture: estimated from crop production in surrounding flatlands, but declining over time to reflect loss in productivity due to erosion from slopes in KYNP
2. Eucalyptus plantations: 12-year investment horizon (rotation), with current costs and prices assumed to persist. Given difficulty in sustaining production due to fragile soils that are poor in nutrients, there are three sustainability scenarios: the plantation is productive for one, two, or three periods.

Results: The benchmark case is presented in Fig. 4.5a. We focus on the decision for the first period, and compare the traditional treatment of uncertainty and irreversibility with the true stochastic dynamic optimum, as explained earlier in the chapter. In both cases, already degraded zone 1, which starts in $M$, is converted to

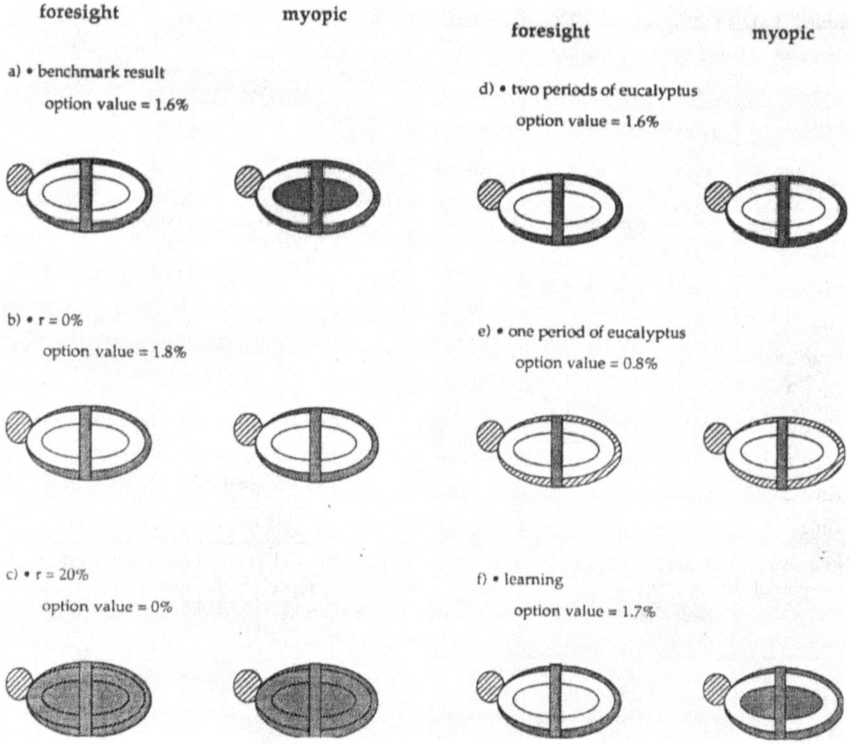


**Fig. 4.5** Khao Yai National Park (KYNP) first period optimal patterns

$D$. In both cases, Zone 2 is preserved. And in both cases, Zone 4 is used for $M$: intensive extractive goods and small scale agriculture.

But Zone 3, which represents approximately half of the total area, is developed in the traditional regime and preserved in an optimal decision. Note that the option value in this case, though only 1.6% of the value of the 3-period sequence in the optimal configuration, leads to a dramatic difference in land use.

Remarks and sensitivity analysis: Option value will be larger, the nearer in time is the uncertainty (remember that it is at least 12 discounted years in the future), the lower the discount rate, and the larger the variation in possible outcomes (here there is very little assumed variation; in a more realistic representation of uncertainty, option value would presumably be still more important).

Panels Fig. 4.5b, c show the effect of the discount rate on both option value and the decision to preserve. At a zero discount rate, option value is a bit larger, and at 20%, option value vanishes as all of the land is developed in both regimes.

Panels Fig. 4.5d, e show the impact of assumptions about sustainability of the development alternative. Assuming the eucalyptus plantation fails after two periods leads to preservation of Zone 3 in the traditional regime. Assuming failure after one period leads to retention of Zone 1 in the intermediate use state in both regimes.

Panels Fig. 4.5f illustrates what happens when the resolution of uncertainty over the benefits in later periods is formulated in a somewhat different but probably more realistic way. Recall that up to this point uncertainty about benefits in a period is resolved at the beginning of the period, and this does not affect uncertainty about future periods. Now suppose the state revealed in one period changes the probability distribution for the next period. In our example, the state revealed in the second period changes the probability of each state in the third period. The probabilities in the benchmark case were: probability of the high state in the third period, $p(h) = 0.25$, and probability of the low state, $p(l) = 0.75$. These probabilities are unaffected by the second period realization. Now let $p(h) = 0.5$ and $p(l) = 0.5$ if the high state is realized in the second period, and $p(h) = 0$ and $p(l) = 1.0$ if the low state is realized. A high state in the second period shifts the distribution for the third state, putting more weight on a high state outcome and less on a low state outcome. A low state in the second period results in movement in the opposite direction: less weight on a high state outcome and more on a low state. In this example the optimal land allocation in the first period is not affected, although the option value is increased by 8%, from 1.6 to 1.7. In another example, with somewhat different benefit numbers, the increase in option value might tip the allocation.

# Chapter 5
# Resources, Growth and Sustainability

**Abstract** This chapter returns to the Malthusian question: are we running out of resources, whether due to population growth or rates of use of both exhaustible and renewable resources. More generally, the question is whether the natural resource base of the economy is adequate to sustain current standards of living. The literature on attempts to estimate measures of resource scarcity at different points in time is reviewed, with some new material, focusing on both physical measures such as reserves, and reserves/production ratios, and economic ones such as costs and prices. The modern statement of the question further broadens the definition of resources used in the production of goods and services (including environmental services as analyzed in the preceding chapter) to include natural capital, contributions of the natural environment such as soils, unpolluted water, and clean air to an inclusive measure of welfare. This in turn leads to the concept of sustainability or sustainable development, first defined in a 1987 report as development that meets the needs of the present without compromising the ability of future generations to meet their own needs. The main thrust of this part of the chapter is to translate this statement into a form suitable for rigorous analysis and ultimately for measurement. This is accomplished first with an intuitive model and then in a more formal one, with much the same results. A very preliminary attempt in the literature at measurement of the sustainability of a number of national economies is reported.

Since the time of Malthus's (1798) analysis of the implications for the human economy of a growing population and limited agricultural land, the relevant resource in pre-industrial England, the concept of what constitutes a (natural) resource has evolved and broadened dramatically. Ricardo, writing a few years later, included as resources minerals on or under the land, and distinguished between higher and lower qualities of both. What was limited was not so much all land as higher quality land, which could yield output at lower cost in Ricardo's formulation. And Mill, the last of the great classical economists, writing around the middle of the 19th century in the early stages of the industrial revolution, considered depletion of mineral deposits but also that stocks of resources like minerals could be expanded by new discoveries, and even that natural resources, including land, can be valuable not only for what can be extracted or produced from them but for their on-site, unmodified features. Since





A. C. Fisher, *Lecture Notes on Resource and Environmental Economics*, The Economics of Non-Market Goods and Resources 16,
https://doi.org/10.1007/978-3-030-48958-8_5

at least the 1970s, features of the natural environment such as clean or unpolluted air and water, in urban and rural as well as in natural or wilderness areas, have been considered as resources subject to "depletion" by the discharge of residuals from production and consumption activities. Still more recently, a stable global climate has come to be considered in much the same sense as a resource.

Whatever the definition of resources, the underlying issue, dating all the way back to Malthus and very much present today, is whether the natural resource base of the economy is adequate to sustain current standards of living, or indeed to support growth. Another way of putting it is: are we running out of resources? And how do we know? And does it matter? We tackle these questions in three sets of lectures, starting with the traditional concern for the continued availability of commercial extractive resources such as metals and fuels. How is scarcity, or prospective availability, best measured? And what have empirical studies shown?

We then broaden the focus in accord with the evolving definition of resources, by including environmental resources, in a brief discussion of the well-known and influential work "The Limits to Growth," first published in 1972 and in a revised edition in 1992. Both are based on a computer simulation of the dynamic interactions among several macro-level variables including (traditional) natural resources, capital, population, agricultural and other output, and environmental pollution. The key question the study seeks to answer is how do assumed resource and environmental limits affect the ability of the model economy to expand to meet the needs of a growing population.

A critical analysis of the Limits model leads to a consideration of what may be called the modern statement of the problem, under the heading of "sustainability." Though one can no doubt find earlier mentions, the concept of sustainability was brought to the attention of the general public, policy analysts and economists by the report of the World Commission on Environment and Development 1987, also known as the Brundtland Commission, after the Chair, sometime Prime Minister of Norway Gro Harlem Brundtland. Sustainable development was defined as: "Development that meets the needs of the present without compromising the ability of future generations to meet their own needs." Our objective will be to translate this statement into a form suitable for measurement, at least in theory. We also consider a preliminary, though promising, attempt at measurement of the sustainability of a number of national economies.

## 5.1 Resource Scarcity

Concern for the continued availability of production materials such as wood, iron and coal, was a prominent feature of the early (1890–1920) Conservation Movement in the U.S. The modern study, focusing on the measurement and projection of scarcity, can be traced to the report of the President's Materials Policy Commission in 1952, also known as the Paley Commission after Chair William Paley. The Commission, formed in 1951 by President Harry Truman, was charged to examine

**Table 5.1** Ratio of reserves to annual production hardly changing over time

| Year | Reserves (billion barrels) | Reserves/ production |
|---|---|---|
| 1955 | 190 | 33 |
| 1960 | 298 | 37 |
| 1965 | 348 | 30 |
| …<br>1972 | 667 | 35 |
| …<br>1980 | 642 | 30 |
| 1985 | 702 | 36 |
| 1990 | 1002 | 45 |
| 1995 | 1000 | 44 |
| 2000 | 1018 | 41 |
| 2005 | 1278 | 47 |
| 2010 | 1362 | 50 |
| 2015 | 1663 | 57 |

*Sources* 1955–1972: *Oil: World Statistics*, Institute of Petroleum Information, cited by Robinson (1975); 1980–2015: Energy Information Administration

prospects for future and especially long run shortages of key industrial materials, prominently metals and fuels which had been heavily depleted during World War II. The report, "Resources for Freedom: Foundations for Growth and Security", recommended the formation of an independent organization to carry on the work. This was accomplished in the same year by the founding of Resources for the Future (RFF), mentioned earlier in connection with the pioneering work on environmental resources by RFF economists in the 1960s. But the original objective was to determine whether stocks of conventional extractive resources on hand would be adequate to meet projected demands.

### 5.1.1 Physical Measures of Scarcity

Typically, news accounts or discussions in popular media of threats to the continued availability of the raw materials for production focus on estimates of reserves, or the ratio of reserves to annual production or consumption, which gives the number of years to exhaustion, for example for world oil, given in Table 5.1 below.

Reserves are defined as the known amount of a (mineral) resource that can be profitably recovered. This means that new technology, new discoveries, higher prices for the resource, all lead to an expansion in reserves, as we see in the case of oil. The relationship between resources and reserves is expressed in a simplified version of the McKelvey diagram in Fig. 5.1, after Vincent McKelvey, former head of the U.S Geological Survey (USGS).

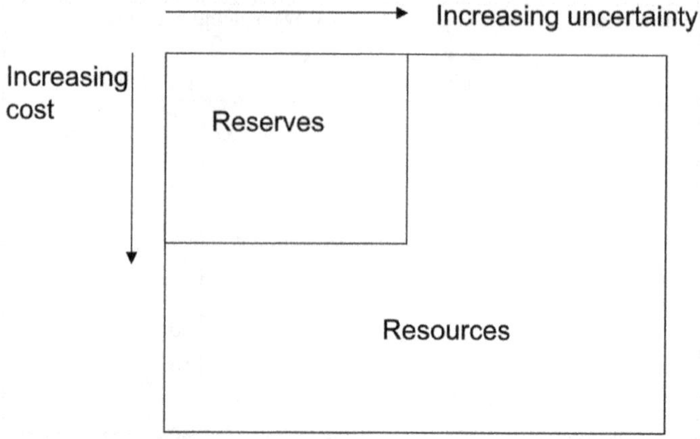


**Fig. 5.1** McKelvey diagram: reserves expand via new technology and new discoveries

Not only do reserves typically increase over time, as in the case of world oil, but the reserves/production ratio is often approximately constant over time, as in the case of oil over the periods 1955–1985 and 1990–2015. In fact, the apparent increase from 1985 to 1990 may be due to a strategically motivated more inclusive classification of material as reserves by some countries. The larger ratios most recently may reflect the classification as reserves of materials that have become economical to extract and produce with new technologies such as hydraulic fracturing, or "fracking". Why the constancy? Reserves can be regarded as inventory. A mining firm—or a government agency—doesn't want to invest too much in costly exploration, though of course some amount will be desirable because it takes time to discover and develop reserves. Thus there is an optimal inventory, which should be approximately constant over time. Observed variations may be explained by the uncertainty of the discovery process and the "lumpiness" of discoveries, as well as changes in technology.

We conclude that estimates of reserves, or reserves/production ratios, are not good indicators of Malthusian resource scarcity, or impending resource exhaustion. What is wanted is an indicator of ultimately recoverable resources. If reserves represent one extreme on the resource spectrum, crustal abundance represents the other. This is the material that exists in minute concentrations in the "average rock" of the earth's crust. Some idea of how these amounts compare with reserves, as cited by Nordhaus (1974) based on USGS data on ratios of reserves:annual consumption and crustal abundance:annual consumption for a number of important minerals, is given in Table 5.2.

The differences in the two ratios are striking. Supplies of the "scarcest" resource, lead, appear sufficient for just 10 years according to the reserve ratio (as of 1974, implying that we should have run out of lead by 1984!), but for 85 million years if we look to crustal abundance. This is certainly an ultimate limit and if it accurately represents resource availability there is no apparent reason for concern. The problem however is that it ignores the costs of finding, developing, extracting, and transporting

**Table 5.2** Crustal abundance and mineralogical thresholds, as of 1974

| Mineral | Reserves/ production | CA (millions)/ production | MT/ production |
|---|---|---|---|
| Aluminum | 23 | 38500 | 68066 |
| Copper | 45 | 242 | 340 |
| Iron | 117 | 1815 | 2657 |
| Lead | 10 | 85 | 162 |
| Molybdenum | 65 | 422 | 630 |
| Phosphorous | 481 | 870 | 1601 |
| Uranium | 50 | 1855 | 8455 |
| Zinc | 21 | 409 | 618 |

*Source* USGS, cited in Nordhaus (1974)

resource deposits, as well as the environmental costs associated with processing such an enormous volume of material. In the case of lead, for example, 77,000 tons of material would be required to produce a ton of the metal, even assuming 100% recovery (Brobst 1979). Note, by the way, that the 1974 calculation is relevant to the point about reserves and crustal abundance. Crustal abundance has not changed, and as we have argued, the reserves:consumption ratio tends to remain steady over time. Of course, annual consumption has increased—as Malthus might have put it, arithmetically—but this would not make a meaningful difference in the comparison of ratios.

It follows that a measure in between these extremes would be more informative. Brobst proposes the "mineralogical threshold", above which a metal element occurs in minerals of its own, rather than as a trace element. The ratio of resources above the mineralogical threshold to annual consumption is given in the third column in Table 5.2. It is calculated as 0.01% of crustal abundance to a depth of one kilometer, which is the amount of copper available in its "own" deposits, i.e., not just as a trace element, to this depth. This is no doubt a more meaningful measure of scarcity than either reserves or crustal abundance, but is of course somewhat arbitrary, and may not be applicable to materials other than copper. More importantly, it represents a very crude attempt to take into account economic considerations, such as cost, to which we now turn.

### *5.1.2 Economic Measures of Scarcity*

#### 5.1.2.1 Cost and Productivity

Consideration of trends in costs of resource extraction, or more generally production, follows directly from the Ricardian model of exploitation of agricultural land and resources in order of increasing costs. The best land and the richest deposits of

**Table 5.3** Labor-Capital input per unit of extractive output (1929 = 100)

| Period | Total extractive | Agriculture | Minerals | Forestry |
|---|---|---|---|---|
| 1870–1900 | 134 | 132 | 210 | 59 |
| 1919 | 122 | 114 | 164 | 106 |
| 1957 | 60 | 61 | 47 | 90 |

*Source* Barnett and Morse (1963, p.8)

minerals are used first, but as population and the economy expand, poorer quality and therefore higher cost land and resources are drawn into production. It follows that a time series of costs for specific resources or categories of resources should show increasing costs, signifying increasing scarcity of good land and rich deposits. Surprisingly, in what remains in my judgment the most influential single work on the subject of resource scarcity, "Scarcity and Growth", by Harold Barnett and Chandler Morse (1963), capping the first ten years of research by economists at RFF, this was not found.

As a side note here, in interpreting their results, the authors suggest a wider definition of costs, social as opposed to the purely private costs that they measure. In particular, the social costs would include environmental disruption of various kinds, which they hypothesize may have in fact been increasing. This notion would go on to be developed by RFF economists and others over the next decade. Two distinct strands can be distinguished. One, the more familiar, studies the causes and consequences of environmental pollution and analyzes alternative approaches to control. The other, with which we have been primarily concerned in the preceding set of lectures, can be called the "new conservation economics", introduced in the seminal paper "Conservation Reconsidered" by Krutilla (1967) and focusing on the fate of the natural environments from which commercial extractive resources are taken.

To return to "Scarcity and Growth", what exactly did Barnett and Morse find? A brief summary of results is given in Table 5.3, showing unit costs for all extractive resource outputs, and separately for agriculture, minerals, and forest products as estimated for the period 1870–1957, encompassing the development of heavy industry in the U.S., with greatly increased demands on resources, especially minerals. The authors present a great wealth of additional detail, including a description of how costs were estimated as a measure of labor and capital inputs per unit of output, where an estimate of capital inputs could be measured (for agriculture and minerals in the aggregate, which together account for 90–95% of extractive output) and labor inputs per unit of output elsewhere (forests and fisheries). This painstaking approach was necessary as there was no available time series on costs, nor is there today for researchers seeking to extend the analysis.

The deceptively simple table captures the essential finding: for agriculture, and especially for minerals, costs have fallen dramatically, contrary to the simple Ricardian hypothesis. Only for the relatively unimportant forestry sector is this trend not observed, and even here costs fell over the second part of the period. In fact, the cost of forest output over the first part of the period appears to follow a classical Ricardian pattern, in which exhaustion of the better and more accessible stands of trees dictated resort to poorer quality, or higher cost, sources.

Here we need to note that the calculation, heroic as it is, omits not only changes over time in costs due to pollution or other environmental degradation, as acknowledged by the authors in discussing their findings, but also in the cost of purchased inputs such as materials. In the case of agriculture, for example, animal feed and fertilizer, formerly produced on-farm, are now largely produced off-farm. Ideally, either the labor and capital used for this purpose in more recent years would be included as an input to farm production or the cost of the purchased input would be included along with the costs of on-farm labor and capital. In fairness to the procedure followed by Barnett and Morse and the authors of subsequent studies, labor and capital no doubt have accounted for a large fraction of the cost of conventional inputs to extractive resource production, so that changes in expenditure on other inputs may not make much difference.

Accepting the results, which despite the qualifications are probably qualitatively correct, and are certainly correct neglecting environmental impacts, where did Ricardo go wrong? The answer is, he didn't, in theory. Nor, for that matter, did Malthus, whose prediction of diminishing returns or, equivalently, increasing costs of agricultural output as additional labor crowded onto a fixed or only marginally increasing stock of land has not been borne out. The difficulty is that the stock of agricultural land in the U.S., much of it high quality ("Ricardian") land, increased dramatically, not marginally, more than doubling over the first part of the study period, from 406 million acres in 1870 to 955 million in 1920.[1] A closer look at American agricultural history offers an explanation for the great expansion. By the mid-1860s significant agricultural settlement had begun on the Great Plains, with heavy settlement in the years following 1880 and continuing until around 1920. The total increase, which comes to approximately 550 million acres, closely matches current (as of 1992) farmland acreage of 512 million on the Great Plains.[2] The small difference presumably can be accounted for by marginal settlement elsewhere.

This could help to account for the decline in costs of production. But as indicated in Table 5.3, the decline accelerated in the second part of the period, accompanied by a smaller increase in farmland of about 150 million acres to 1960 and indeed a decrease back to 1920 levels by 1998, so further explanation must lie elsewhere. Recall that for both Malthus and Ricardo, technology was essentially fixed. A one-time improvement in "the arts of cultivation" could temporarily boost output, or reduce costs, but diminishing returns or increasing costs would inevitably set in as population continued to grow and crowd onto the land. This was perhaps a reasonable scenario in the late 18th and early 19th centuries, but by around the middle of the 19th century the industrial revolution was having a profound impact on the agricultural economy, manifested in a continuing stream of technical innovations including substitution of mechanical power, embodied in a great variety of new machines, for

[1] https://www.agclassroom.org.

[2] Encyclopedia of the Great Plains, online at http://plainshumanities.unl.edu/encyclopedia/.

**Table 5.4** Cost-reducing technical innovations for corn production in the midwest and great plains (1850–1975)

| Year | Hours of labor to produce 100 bushels of corn | Technology |
|---|---|---|
| 1850 | 75–90 | Walking plow |
| | | Harrow |
| | | Hand planting |
| 1890 | 35–40 | 2-bottom gang plow |
| | | Disk |
| | | Peg-tooth harrow |
| | | 2-row planter |
| 1930 | 15–20 | 2-bottom gang plow |
| | | 7-foot tandem disk |
| | | 4-section harrow |
| | | 2-row planter, cultivator, and picker |
| 1945 | 10–14 | Tractor |
| | | 3-bottom plow |
| | | 10-foot tandem disk |
| | | 4-section harrow |
| | | 4-row planter and cultivator |
| | | 2-row picker |
| 1975 | 3.5 | Tractor |
| | | 5-bottom gang plow |
| | | 20-foot tandem disk, planter, and herbicide applicator |
| | | 12-foot self-propelled combine harvester, and trucks |

*Source* Historical Timeline—Farm Machinery and Technology, Agriculture in the Classroom (https://www.agclassroom.org)

human and animal effort; discovery and introduction of improved breeds, including hybrid crops and crossbred livestock; development of new and powerful fertilizers, and so on.

One example that necessarily captures just a small part of this ongoing process can illustrate its importance. The numbers, displayed in Tables 5.4 and 5.5, tell the story for the major grain crops of the Midwest and Great Plains, corn and wheat.

Impressive as these trends are, there are a couple of caveats, similar to those noted with respect to the results in Scarcity and Growth. First, labor per unit of output is only a partial measure of cost. Still, although the new capital has not been costless, technical change in agriculture has saved more in labor per unit than it has cost in capital, taking into account that it has also reduced cost per unit by decreasing the pressure on the land required to provide calories for work animals (Barnett and Morse 1963, p.197). And second, in addition to capital, other inputs such as materials

**Table 5.5** Cost-reducing technical innovations for wheat production in the midwest and great plains (1850–1975)

| Year | Hours of labor to produce 100 bushels of wheat | Technology |
|---|---|---|
| 1890 | 40–50 | Gang plow |
| | | Seeder |
| | | Harrow |
| | | Binder |
| | | Thresher |
| | | Wagons |
| | | Horses |
| 1930 | 15–20 | Tractor |
| | | 3-bottom gang plow |
| | | 10-foot tandem disk |
| | | Harrow |
| | | 12-foot combine harvester, and trucks |
| 1955 | 6.5 | Tractor |
| | | 10-foot plow |
| | | 12-foot row weeder |
| | | Harrow |
| | | 14-foot drill |
| | | Self-propelled combine harvester, and trucks |
| 1975 | 3.75 | Tractor |
| | | 30-foot sweep disk |
| | | 27-foot drill |
| | | 22-foot self-propelled combine harvester, and trucks |

*Source* Historical Timeline—Farm Machinery and Technology, Agriculture in the Classroom (https://www.agclassroom.org)

purchased off-farm formerly produced on-farm need to be accounted for to get a complete picture of changes over time in the cost of agricultural output.

With respect to minerals, as shown in Table 5.3 costs of production experienced an even steeper decline than in agriculture. As suggested in the McKelvey diagram, reserves, the known amounts of a resource that can be profitably produced, can be (and have been, as indicated for world oil in Table 5.1) greatly expanded by means of new discoveries and more importantly new techniques for finding, extracting, and converting resources. Thus additional resources have been found due to advances in geological knowledge and search techniques and converted into reserves by advances in the technology of extraction and conversion. A striking example given by Barnett and Morse is that porphyries of less than 1% copper content yield copper concentrate at a lower unit cost per ton of copper content than former high grade ores, because

**Table 5.6** Annual growth rate in unit extraction costs, 1870–1957 and 1958–1966 (agriculture and minerals, labor plus capital inputs) and 1870–1957 and 1958–1970 (forests and commercial fisheries, labor inputs only)

| Annual growth rate (percent) | 1870–1957 | 1958–1966 | 1958–1970 |
|---|---|---|---|
| All extractive | −1.4 | −1.7 | |
| All agriculture | −1.6 | −1.8 | |
| All minerals | −1.7 | −2.1 | |
| Forestry | +1.0 | | −2.9 |
| Commercial fishing | −0.74 | | +1.6 |

*Source* Johnson et al. (1980, Tables II and III)

of advances in open-pit, earthmoving, and digging techniques and the discovery of selective flotation techniques to concentrate the copper content of pulverized ore (p.199).

A natural question at this point is, what happened next? That is, have the cost trends continued beyond 1957, or reversed to indicate growing scarcity? Johnson et al. (1980) update the record to 1970 (more precisely in some cases a little earlier or a little later, depending on the resource or resource aggregate), providing a complete century of data and also conduct an econometric analysis of the time trends, regressing costs on time, along with a dummy variable which takes the value of zero for the years 1870–1957 and one for the post-1957 period. There are other elements to the analysis which need not concern us here. The short answer to the question about what happened is that the decline in unit costs continued uninterrupted. However, use of the dummy variables makes possible a more informative answer, comparing the behavior of costs over the two distinct periods.

Results are presented in Table 5.6, taken from more detailed tables in Johnson, Bell and Bennett. I have included results for forests and fisheries, which display more varied and interesting behavior, but note that for these two sectors only labor inputs have been measured, so the usual caveats apply.

As the table shows, unit extraction costs continued to decline in the post-1957 period, and indeed at an increasing rate across the categories of all extractive resources, all agriculture, and all minerals. The reversal of the classic Ricardian trend in forestry may be explained by the heavy substitution of other structural materials for lumber as a result of technical change in these industries and the rising cost of lumber, in turn reducing the pressure on forest resources in the later period, in which output remained relatively constant while labor costs decreased. We should note that the results for forestry would be weakened by inclusion of capital costs as capital has increasingly substituted for labor. For fisheries, the pattern is just the opposite, with declining costs early on followed by rising costs, likely due to over-exploitation of ocean fisheries, a classic common-property resource. While this is a problem, it's not a problem of Malthusian or Ricardian scarcity.

An update on cost trends by Hall and Hall (1984), based on similarly specified regressions comparing trends in the 1960s to trends in the 1970s shows that for the category of all agriculture, unit production costs continued to decrease in the 1970s, though less rapidly than in the 1960s. On the other hand, unit costs for oil and natural gas (combined) switched from decreasing to increasing. The authors test for the influence of the OPEC oil price shock in 1974 with a dummy variable that takes the value of zero for the years prior to 1974 and one from 1974 on and find that it was statistically insignificant. This may seem surprising, but remember that we're talking about costs, in the form of labor and capital inputs to production, and not prices. The same qualitative results are obtained for coal from the earlier period to the later, i.e. unit costs first falling, then rising, and an insignificant effect of the 1974 oil price shock.

What about more recent trends in the behavior of costs of extractive output? We have some data, primarily for agriculture, on changes over certain periods in productivity, essentially the inverse of the measure of cost per unit of output we have been discussing: the ratio of total output to total inputs (labor, capital, materials). Note that this can also be expressed in terms of a single input, typically labor, or the combination of labor and capital, though as for the cost trends this is clearly less satisfactory than a measure of total factor productivity. An increase in productivity is usually understood as the increase in output not accounted for by increases in inputs. For example, if agricultural output increases from one year to the next by 2% and the use of inputs by 1%, then productivity is said to have increased by 1%.

The Economic Research Service (ERS) of the U.S. Department of Agriculture (USDA) has estimated that total factor productivity in U.S. agriculture increased at an annual average rate of 1.45% between 1948 and 2013, and was almost solely responsible for the growth in output over the period. Output growth averaged 1.50% per year, and inputs in total increased by an average of only 0.05% per year, where the large decline in labor and smaller decline in capital were offset by increases in intermediate goods. This is shown in Table 5.7, where growth rates in output, inputs and productivity are separately broken down by subperiods. Note that use of capital and labor has been declining since 1980, so the increase in total outputs has been mostly driven by increases in productivity, and to a lesser extent, increases in intermediate goods. Put differently, the cost per unit of agricultural output continued to decrease over this period.

For minerals, data on recent cost trends (that I have been able to find) is extremely limited, and as for agriculture concerns changes in productivity, Specifically, and courtesy of The Mining Association of Canada, we have that annual average productivity growth in mineral extraction over the period 1997–2006 was 1.8%. Presumably a similar number would hold for the U.S.

In the face of this somewhat discouraging evidence (with the exception of the Hall and Hall results on the cost of extraction for oil and natural gas in the 1970s), a resource pessimist might argue that it proves nothing about the future. We might be coming to the end of a "resource plateau", to the end of a period of technical change that has kept costs from rising for many decades. It's also worth noting that there is a limit to this process; costs can't fall below zero. Is there a measure that signals impending exhaustion of a resource even in the face of technical advances that keep down and in fact reduce costs? The answer, as suggested by the theory

**Table 5.7** Average annual percentage growth in output, inputs, and productivity in U.S. agriculture, 1948–2013

| Period | Total outputs | TFP | Total inputs | Capital | Labor | Intermediate goods |
|---|---|---|---|---|---|---|
| 1948–2013 | 1.50 | 1.45 | 0.05 | −0.18 | −2.18 | 1.24 |
| 1950–1959 | 1.88 | 1.41 | 0.47 | 0.35 | −3.42 | 2.53 |
| 1960–1969 | 1.34 | 1.39 | −0.05 | 0.26 | −3.08 | 1.22 |
| 1970–1979 | 2.62 | 1.38 | 1.24 | 0.34 | −0.93 | 2.63 |
| 1980–1989 | 1.11 | 2.98 | −1.87 | −2.07 | −1.86 | −1.2 |
| 1990–1999 | 1.84 | 1.30 | 0.54 | −0.77 | −0.62 | 1.72 |
| 2000–2010 | 0.61 | 0.99 | −0.38 | −0.19 | −1.59 | 0.23 |

*Source* USDA-ERS *Agricultural Productivity in the U.S.* Available at: https://www.ers.usda.gov/data-products/agricultural-productivity-in-the-us/ (Accessed 13 Mar 2017)

of optimal depletion, is of course yes. In fact, there are two: resource price and resource royalty, or rent. The royalty is particularly relevant for exhaustible resources, but not agriculture, in principle a sustainable, renewable resource unless soils are significantly or even irreversibly depleted. Thus we might expect agricultural price behavior over time to reflect movement in costs, with however greater variability due to other influences on prices, such as changes in macroeconomic activity.

There are other, less abstract reasons for going to a price measure. Unlike costs, which as we have seen are not published or otherwise available, rather need to be constructed in painstaking and even heroic fashion, prices are relatively available, though tying together different sources from different periods to produce a series of sufficient length to be interesting can be challenging. Another advantage of prices is that they necessarily include purchased intermediate inputs, not just labor, or labor plus capital. A qualification is of course that they don't include unpriced environmental resources which may have been in some fashion consumed in production of the resource, in the absence of policy to internalize externalities. And as Barnett and Morse point out, the price measure is statistically independent of the cost measure, so it will be interesting to see the extent to which it comes to qualitatively similar results despite the differences in concept and construction just noted.

#### 5.1.2.2 Price

If one is interested in scarcity of the "pure" resource, unmixed with human labor or other productive factors, the royalty, the value of a unit of the resource "in the ground" is the appropriate measure. If, on the other hand, one is interested in the sum of sacrifices made to obtain a unit of the resource, then price, which includes both cost and royalty, is relevant. Because the royalty is ordinarily not observable (though it might be inferred at a point in time from bids on mineral or timber leases, which reflect the bidders' estimates of the value of the resource in the ground net of extraction costs) most empirical work in this area has focused on price. Another landmark early RFF study, by Neil Potter and Francis Christy (1962), developed time series for prices, output, employment, and trade in natural resource commodities over the period 1870–1957 for the U.S. The output and employment series were in fact used by Barnett and Morse, along with their estimate of capital inputs where possible, to construct their cost estimates. A chart of the (deflated) price time series for all extractive resources and separately for agriculture, forests, fisheries and minerals displayed in Barnett and Morse shows a great deal of fluctuation over time in each but no clear trend for all extractive resources, agriculture and fisheries, along with a modest decline in mineral prices and an equally modest rise in forest product prices.

Given the heavy emphasis in the subsequent (post-1970) literature—scientific, popular and economic—on prospects for sustainability and growth in the U.S. and world economies in the face of resource limitations, we have constructed resource commodity price series along the lines of those in Potter and Christy, but over a much longer period, up through 2015 and in some cases all the way back to 1841, capturing the industrial history of the U.S from the very beginning right up to the present. Further, the series are disaggregated to highlight the trends in key resources such as the energy minerals.

Before proceeding we need to ask why price time series are informative in studying resource scarcity, but not, as I argued earlier, in testing empirically the validity of the theory of optimal depletion of exhaustible resources. Recall that the difficulty in the latter case is that the series reflect many things beyond the optimality calculation, chief among them technical change in the finding, extraction, and conversion of resources. But in assessing scarcity these are just the things we are looking for: what do they tell us about prospects for ultimate exhaustion, or sustainability of a resource-based economy? Other influences on prices, such as changes in aggregate demand (recession or depression) and changes in market structure (success or failure of a resource cartel like OPEC in restricting supply and raising prices) tend to be transitory and thus not relevant in a search for long run trends. They will of course be a source of fluctuations, so that we want the series to be as long as possible, to enable us to put transitory effects in their place and not allow them to influence unduly any inferences we can make about prospects for scarcity in the very long run.

The next three figures show time series of (real) prices of major agricultural, non-fossil fuel mineral and fossil fuel commodities from 1841 to 2015. Not all data series span the full range. The series are transformed such that the "price" of each resource is set at 100 for 2015, and all earlier prices are relative. Thus for example as shown in Fig. 5.2(a), the prices of most agricultural commodities spiked in the years 1973–74, coincident with the first oil price shock, to levels between 200 and 400, or from twice to four times the levels in 2015, then fell (though with another, smaller, spike coincident with the second oil price shock in 1979–80) to well below (a little over half) 2015 levels until beginning to rise in around 2003 as part of the "commodities boom", in turn cut short by the Great Recession in 2008.

**Fig. 5.2** Commodity price indices

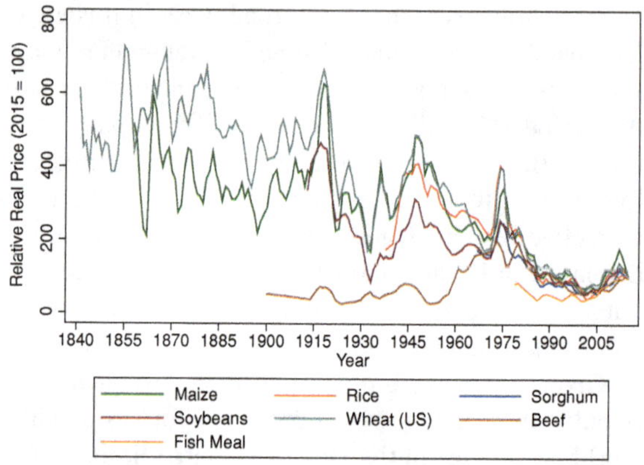


(a) Agricultural Products: 1840–2015

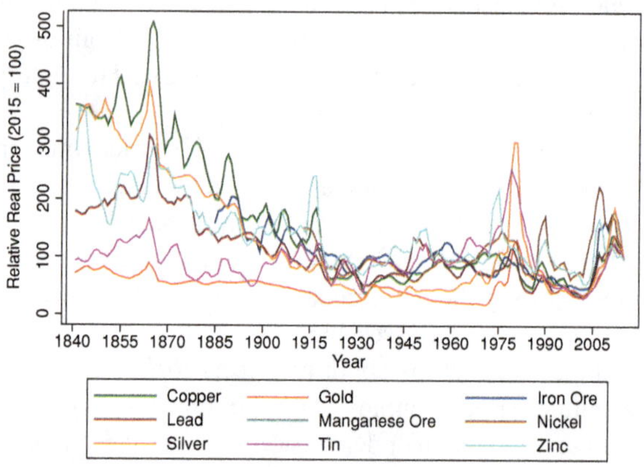


(b) Minerals: 1840–2015

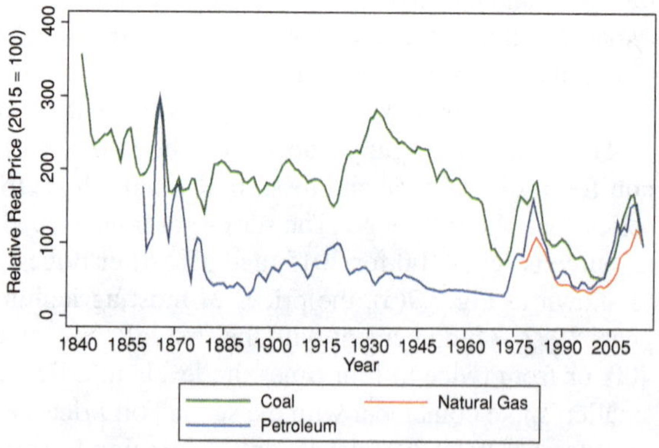


(c) Fossil fuel price indices: 1840–2015

These data extend the price series data of (Krautkraemer 2005) to 2015. The data were downloaded on 6 September, 2016 from the Global Financial Database, provided by Global Financial Data, a firm that collates a variety of different government, academic, and commercial data sources to create consistent, reliable data series. Since all data series used nominal prices in USD, real prices were generated by deflating the nominal prices using the US consumer price index.

In addition to setting 2015 as the base year for indexing, two other changes were made to improve the legibility of the graphs. First, three years of petroleum data (1859–1861) were excluded from the series. For these years the index was an order of magnitude higher than all subsequent years (1959 = 1235; 1960 = 705; 1961 = 472), thus diminishing the readability of the graph. The index was particularly high during these years due to the high costs of extraction in the first years of commercial petroleum production, after which there was a rapid reduction in costs as technology and know-how improved. Second, the aluminum series was removed from the minerals series. For the initial 19 years of the series (1900–1918) the index was an order of magnitude greater than the other commodities. It began from a substantially higher level than the other commodities probably due to the initial high cost of processing bauxite with electricity.

More broadly, two main conclusions can be drawn from "eyeballing" the data. First, there is a great deal of fluctuation in each of the series, but in each case likely due to fluctuations in global macroeconomic activity (2003–2008 and post-2008) or dramatic but short-run changes in supply, as in the cut in OPEC output and associated jump in oil prices in 1973–74, which in turn affected the production costs and prices of other resource commodities. The second main conclusion is that, for all of the fluctuation, for both agricultural and mineral commodities the trend in prices is broadly down, from 1841 or whenever a given series starts to the present. An interesting difference between the two is that, for agriculture, most of the fall in prices is post-1915, with of course some breaks as just noted, whereas for several of the minerals most of the fall takes place in the earlier period, roughly 1841–1915.

#### 5.1.2.3 Royalty

Although I won't have much to say about empirical resource royalties as these are not ordinarily observable, there is a novel way of getting at them in one important case, for oil in the period from the end of the Second World War (1946) to 1971. This is an interesting period as it immediately precedes the oil price shock. To derive an estimate of the oil royalty we need to go back to the theory of optimal depletion in Sect. 2.5.2 and add a new element: exploration, the purposeful addition to the stock of a resource, represented by the cost function, $\phi(z_t)$, where $z_t$ represents new finds, measured in units of the resource.

An obvious objection to proceeding in this way is that the outcome of exploration is presumably uncertain. This is true, but although introducing uncertainty considerably complicates the formulation of the problem, the analytics, and the results, Devarajan and Fisher (1982, JPE)) show that the logic of the solution we shall obtain in the simple deterministic case comes through, albeit in a less tidy fashion.

The firm's problem is to maximize the present value of profits from the resource, where profits are net of the costs of both extraction and exploration. In symbols, this is

$$\max_{\{q_t\},\{z_t\}} \int_0^T [p\,q_t - c(q_t, x_t) - \phi(z_t)]e^{-rt}\,dt$$

The constraint equation is also affected. The change in the resource stock is equal to the difference between discoveries and depletion. With no exploration the stock was monotonically decreasing. Now it may actually increase, if more is found in any one period or interval of time than is extracted. Something like this appears to have happened for many resources, as indicated in our discussion of the behavior of available stocks over time, for example for petroleum.

In symbols, the constraint equation is

$$\frac{dx_t}{dt} = z_t - q_t$$

The firm now has two variables to control, $q_t$ and $z_t$. In addition to extraction $q_t$, it must decide on a target level of new finds, $z_t$. The new finds hold down extraction costs through their influence on the stock, $x_t$. On the other hand, they are not free, so the firm must balance the benefit of reduced extraction cost against the cost of exploration. There is an optimality condition for each control. The condition for $q_t$ is just the same as in the case of no exploration. The condition for $z_t$ is

$$-\frac{d\phi}{dz_t} + \mu_t = 0$$

where $\mu_t$ is the co-state variable attached to the constraint equation.

This tells us that the royalty, or shadow price of a unit in the stock, is just equal to the cost of finding another to replace it in a competitive equilibrium. Previously the royalty was interpreted as the benefit of having an additional unit in the stock. Where a unit can be added, the benefit is just balanced by the cost of adding it.

This is a potentially useful result because it provides a measure of the royalty, in theory a leading indicator of future shortages and price rises. If the royalty is equated to the marginal cost of exploration in a competitive equilibrium, we can examine data on exploration costs to shed light on the behavior of the ordinarily unobservable but theoretically desirable royalty measure of scarcity. A little further on we shall carry out an illustrative exercise of this type using some data on average costs of oil exploration in the years leading up to the oil price shocks of the early 1970s. I say illustrative because the theory specifies marginal costs, not average, and the way to get these requires, first, estimation of an exploration cost function for each period in the study and then differentiation to get marginal costs. To my knowledge, this has not been done, and cannot be with the data we'll look at.

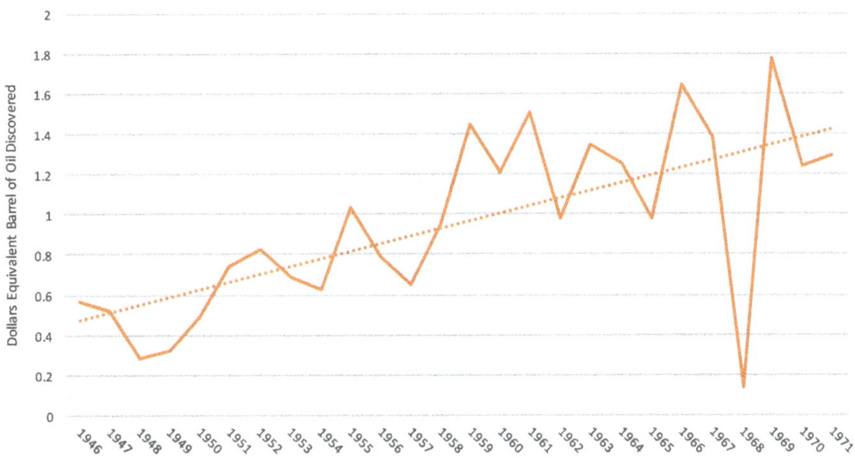


**Fig. 5.3** Average real exploration costs for U.S. oil and gas, 1946–71. *Source* Fisher (1981, p.109)

But first, there is one more theoretical issue that needs to be considered. An exploration cost function should probably include as an argument the sum of previous finds. Just as the cost of extraction is affected by cumulative extraction, the cost of exploration may be affected by cumulative finds. Altering the cost function in this fashion leads, as is easily verified, to an adjustment in the third equation above. The royalty is now equated to an adjusted marginal exploration cost, where the adjustment is a term that can be interpreted as (minus) the shadow price of a unit added to the stock of cumulative finds.

It is difficult to say whether or not this is important, or even in what direction it cuts. If the better deposits are found first, the shadow price will be negative and the royalty will exceed the unadjusted marginal exploration cost. If, on the other hand, early finds provide information that will reduce the cost of future finds, the shadow price will be positive, at least initially. Note the difference from the case of cumulative extraction, in which the affect of greater cumulative extraction, or a smaller remaining stock, on the extraction cost function is unambiguously positive, i.e. the larger cumulative extraction or the smaller the remaining stock, the higher the cost of current extraction.

Now let's take a look at the data on average costs of oil exploration, the admittedly far from perfect measure of the royalty.

Figure 5.3 shows the average real exploration costs for US oil and gas from 1946 to 1971. Costs are measured in dollars equivalent barrel of oil discovered (average 1947–9 dollars). To combine oil and gas discoveries, physical units were aggregated on the basis of market value. The linear trendline has a slope of 0.039, i.e. the average price increase is 3.8 cents per year; the outlier in 1968 has a substantial effect on the slope of this trend. Across this period prices trended upward, increasing 3.8 cents per year on average, an approximately 1.7% yearly increase. There was a substantial fall in the real price in 1968, caused by the discovery of the great Prudhoe Bay,

Alaska oil field, holding about 10 billion barrels of producible oil, or roughly a third of total U.S. reserves at the time. The inclusion of 1968 has a substantial effect on the long-term trend. With 1968 excluded, the yearly average price increase rises to 4.7 cents per year, an approximately 2.4% yearly increase.

What this suggests is that, during a period characterized by low and falling prices, as shown in Fig. 5.2, oil royalties were rising (again, if the exploration cost estimates can be accepted as shedding some light on the behavior of the royalties). A leading indicator?

#### 5.1.2.4 A Brief Summary of Our Discussion of Measures of Resource Scarcity

Reserves and reserves/production ratios are not useful as indicators of future scarcity, though they are commonly used for this purpose, as they measure the amounts of a resource that it is currently profitable to extract, not what may ultimately be produced. Crustal abundance is also not useful; it is much too loose an upper bound. The "mineralogical threshold", above which a metal element occurs in minerals of its own, rather than as a trace element, represents a middle ground in physical measures of scarcity, but is a crude attempt to capture relevant economic considerations such as cost and willingness to pay.

Time series of costs convey useful information on the relevant economic dimension; scarcity is after all an economic concept. But they may fail to capture the notion of anticipated future scarcity, which could be significant. Further, they are not available in published or other form, rather must be constructed in painstaking fashion from what information concerning labor, capital and other inputs in a given resource industry can be developed. Prices are relatively available and necessarily include purchased intermediate inputs, not just labor, or labor plus capital, but a time series of prices may reflect influences unrelated to fundamental scarcity, such as formation and collapse of resource cartels and fluctuations in macroeconomic activity. Assuming these can be identified, time series of resource prices can be good indicators of trends in scarcity, including anticipated future scarcity (through the royalty component of price). Note however that such series are not useful for the purpose of testing simple optimal depletion theory (because of shocks to the system such as new discoveries, technical change, and again, fluctuations in macroeconomic activity).

## 5.2 Limits to Growth

The main and well known result of the simulations in Limits to Growth (LTG) (Meadows et al. 1972, 1992) is that per capita consumption is sharply and inevitably reduced, within a couple of decades in the base case, in several more decades under different assumptions, for example about the initial stock of resources or about the amounts of resources or pollution involved in production of a given output.

Although the model underlying the simulations is complex and involves many assumptions and relationships, following the review by Nordhaus (1973b) we can understand the results by focusing on just two key equations:

$$(1) \qquad c_t = \frac{\alpha\, R_t\, K_t}{P_t}$$

where $c$ = per capita consumption, $R$ = nonrenewable resources, $K$ = capital, $P$ = population or labor, and

$$(2) \qquad R_t = R_{t-1} - \theta(c_t)P_t$$

The parameter $\theta(c_t)$ is a resource use coefficient, e.g. 10 barrels of oil per capita, and is increasing with per capita consumption, $\theta'(c_t) > 0$. Thus $R$ must be declining at an increasing rate with increases in population or per capita consumption; no substitution, and no technical change, as might for example be specified $\theta(c_t, t)$ with $\frac{\partial\theta}{\partial t} < 0$.

As resources are depleted (flow $\Delta R$), and population $P$ is increasing, according to equation (1) the economy needs ever more capital $K$ to maintain per capita consumption c, i.e., an increase in $\Delta R$ implies an increase in $K$. This is the fundamental difficulty with the Limits analysis: no matter how large the initial stock of resources, it is inevitably depleted to the point where the economy can no longer sustain a constant level of consumption, much less increase the level.

Alternatively, suppose, following Nordhaus (1973b, 1992), and received economic theory and empirical findings, we specify a production function that allows for (limited) substitution and technical change:

$$Q = k(1+h)^t \left[P^{\beta_1} K^{\beta_2} (-\Delta R)^{\beta_3} L^{\beta_4}\right]^{1-b} \Pi^b$$

where $h$ is the rate of technical change, equal to 0.025 in this simulation, $L$ = land, and $\Pi$ = pollution. With this specification (nested Cobb–Douglas) of production relations, there is no collapse in per capita consumption. Is it plausible? Yes, up to a point.

There is much empirical evidence and a theoretical basis for a Cobb–Douglas production function with continuing technical change, as Nordhaus specifies, for the conventional inputs $P$, $K$, $-\Delta R$, and $L$. Further, we can certainly imagine substitution between pollution and conventional inputs, by using these inputs to capture and recycle or transform the residuals from production, producing a given level of output with less pollution but more of the other inputs.

But this is not the only possibility. Let us rewrite $Q$ as

$$Q = k(1+h)^t \left[P^{\beta_1} K^{\beta_2} (-\Delta R_E)^{\beta_3} L^{\beta_4}\right]^{1-b} (-\Delta R_N)^b$$

where $R_E =$ conventional, extractive resources, $R_N =$ natural capital (non-extractive). Now the question is, can $K$ substitute for a loss of natural capital ($-\Delta R_N$) in the same way it can for loss of extractive resources ($-\Delta R_E$)? This formulation suggests a less negative view of the Limits findings. They may be pointing in the right direction, if for the wrong reasons. This brings us to the subject of sustainability, which as I have suggested can be considered the modern statement of the Malthusian problem.

## 5.3 Sustainability: An Intuitive Approach

Suppose, following Solow (1974b), two economies produce the same conventionally measured national income, NDP (Net Domestic Product, allowing for depreciation of capital), but one of them uses a large fraction of its $R$, which for the time being we'll let stand for nonrenewable extractive resources, whereas the other one conserves its stock of $R$. It seems obvious that the second economy is in some sense doing better, though this is not reflected in measured NDP.

There is in principle a way to adjust the national income accounts to reflect the change in the stock of resources, but it is difficult in practice, and much more difficult, as we shall see, when we expand the definition of R to include non-extractive natural capital. Further, the adjustment turns out to be essential in defining and measuring sustainability.

NDP is adjusted, or corrected, to reflect the depletion of $R$ (or net depletion, if there is also some discovery). How is the physical depletion evaluated, to make it commensurate with NDP? By its shadow price, or scarcity value, which is of course just the resource royalty, a measure of the value of the resource in the ground based on what it will (optimally) contribute to future production.

The adjustment thus $= \lambda(-\Delta R)$ where $\lambda$ is the royalty. The royalty is not ordinarily observed, though in a competitive equilibrium it will be equal to the difference between price and marginal cost, and as noted earlier might also be inferred from bids on mineral or timber leases, which reflect the bidders' estimates of the value of the resource in the ground, net of extraction costs, and perhaps also from exploration costs, if one wants to push the theory.

What is the link to sustainability? Following the early and influential statement in the 1987 report of the Brundtland Commission. Let us define sustainability as the ability of each future generation to be as well of as its predecessors, that is, to be endowed with whatever it takes to produce a given standard of living (which might be defined by consumption, or more broadly by utility), say the present generation's. What is the nature of the endowment? Resources, $R$, of course, but also capital, $K$, including stocks of human capital. A sustainable path is one that replaces whatever it takes from its inherited natural and produced endowment with more of the same, if possible, or other kinds of capital that can yield consumption or utility.

How much replacement capital is needed? Clearly, enough to maintain the aggregate social capital stock. So the condition for sustainability is: invest aggregate resource rents or royalties in reproducible capital:

$$\lambda(-\Delta R) = \mu \Delta K$$

where $\lambda$ is the resource royalty and $\mu$ is the shadow price of capital. This condition is known as the Hartwick Rule, after Hartwick (1977), who first derived it on the basis of a number of simplifying assumptions. Specifically, he showed that a constant level of per capita consumption would be generated by investment of aggregate royalties from nonrenewable resources in the formation of capital, where resources and capital are the only two inputs to a Cobb–Douglas production function for a single consumption good output. It has since been extended and modified to apply more generally. Note that there is a very strong assumption here, namely that it is possible to substitute capital for resources.

Consider a vector of capital and resource stocks, where we now distinguish different types of capital and resources: $K_1, K_2, \ldots, R_1, R_2, \ldots$ The value of the aggregate capital stock, including stocks of resources, which we might now expand to refer to all forms of natural capital, both extractive and non-extractive, as in the preceding discussion, is $\mu_1 K_1 + \mu_2 K_2 + \cdots + \lambda_1 R_1 + \lambda_2 R_2 + \cdots$ where $\mu_1$ is the shadow price of $K_1$, $\lambda_1$ is the shadow price of $R_1$, and so on.

The substitutability assumption is that a depletion of, say, $R_1$, where $R_1$ is a conventional extractive resource, evaluated as $\lambda_1(-\Delta R_1)$, can be compensated by investment in say $K_1$, such that

$$-\lambda_1 \Delta R_1 = \mu_1 \Delta K_1$$

Suppose this is not true for all $R_i$, i.e., suppose some elements of the natural capital stock can not be substituted. Then either sustainability is not possible, or sustainability requires that the economy maintain a stock, perhaps the existing stock, of the non-substitutable natural capital.

This is called "strong sustainability." Maintaining the aggregate capital stock (natural and produced) via replacement investment in producible capital, is called "weak sustainability." In theory, and in practice, we might expect a mix of both in pursuit of a sustainability objective.

An important point to note here in connection with possibilities for substitution is that substitution can be accomplished by changing the capital-intensity mix of the consumption good, or perhaps more intuitively, substituting in consumption, away from goods that require the particular non-substitutable natural capital input and in favor of goods that do not require this input. Two issues remain: Which elements of $R$ (which $R_i$) in fact fall into the category of non-substitutable? This is a crucial empirical question if sustainability is a policy objective. At what level should the non-substitutable stocks be maintained? Must it be the current level? The maximum sustainable level? Some minimum level below which essential services can no longer be provided, or other adverse consequences follow? We thus conclude by focusing on empirical issues of substitutability—and technical change, which can obviously affect prospects for substitution—much as in the discussions of resource scarcity and limits to growth.

### 5.3.1 *Sustainability: A Rigorous Formulation and Preliminary Empirical Results*

A rigorous theoretical formulation and preliminary empirical approach to sustainability is provided by Arrow et al. (2004). As Arrow et al. note, several interpretations of sustainability are compatible with this definition. Their approach, which leads to a criterion for sustainability essentially the same as the one we have already put forward, is as follows.

Define intertemporal social welfare $V$ at time $t$ as the present discounted value of the flow of utility from consumption from the present to infinity, discounted using the constant rate $\delta$. In symbols,

$$V_t = \int_t^{\infty} U[C_S]e^{-\delta(s-t)}ds$$

Note that $V_t$ is a function of the stock of all of society's productive assets $K_t$ (human capital, physical capital, natural capital, etc) at time $t$, and also of time $t$. In the case where $V$ is stationary ($t$ itself does not directly influence $V$), $V_t$ is a function of $K_t$: $V_t = V(K_t)$.

Now we can define sustainability in terms of $V$ and $K$. The sustainability criterion is satisfied at time $t$ if $\frac{dV_t}{dt} \geq 0$. Let $K_{it}$ denote the stock of the $i$'th capital good at date $t$. By the chain rule of differentiation,

$$\frac{dV_t}{dt} = \sum_i \left(\frac{\partial V_t}{\partial K_{it}}\right)\left(\frac{dK_{it}}{dt}\right) = \sum_i p_{it} I_{it}$$

where $\partial$ is the symbol for partial derivative where $p_t = \frac{\partial V}{\partial K_{it}}$ is the shadow accounting price of $K_t$ and $I_t = \frac{dK_{it}}{dt}$ is the change in $K_t$ over the short interval $dt$. Then

$$\frac{dK_i t}{dt} \geq 0 \rightarrow \sum_i p_{it} I_{it} \geq 0$$

where $\sum_i p_{it} I_{it}$ represents "genuine investment" in Arrow et al.

Several features of this criterion for sustainability deserve emphasis. First of all, even if a consumption path were to satisfy the criterion today and at all future dates, it would not guarantee that utility U at each future date will be as high as it is today. Secondly, the criterion does not identify a unique consumption path; it could in principle be met by many paths. Third, if exhaustible resources are sufficiently important in production and consumption, then it is conceivable that no sustainable development program exists. Fourth, satisfying the sustainability criterion today does not guarantee that it will be satisfied at all future times; a given consumption path may imply a rising V over the interval from now to the next period, but a falling V when it is evaluated over some future interval of time. And, finally, there is no presumption that a sustainable consumption path is optimal in the sense of maximizing $V$.

**Table 5.8** Genuine investment and components as a percentage of GDP

| Country | Domestic net investment | Education expenditure | Natural Resource Depletion | | | | Genuine investment |
|---|---|---|---|---|---|---|---|
| | | | Damage from $CO_2$ emissions | Energy depletion | Mineral depletion | Net forest depletion | |
| Bangladesh 1973–2001 | 7.89 | 1.53 | 0.25 | 0.61 | 0.00 | 1.41 | 7.14 |
| India 1970–2001 | 11.74 | 3.29 | 1.17 | 2.89 | 0.46 | 1.05 | 9.47 |
| Nepal 1970–2001 | 14.82 | 2.65 | 0.20 | 0.00 | 0.30 | 3.67 | 13.31 |
| Pakistan 1970–2001 | 10.92 | 2.02 | 0.75 | 2.60 | 0.00 | 0.84 | 8.75 |
| China 1982–2001 (without 1994) | 30.06 | 1.96 | 2.48 | 6.11 | 0.50 | 0.22 | 22.72 |
| Sub-Saharan Africa 1974–82; 1986–2001 | 3.49 | 4.78 | 0.81 | 7.31 | 1.71 | 0.52 | −2.09 |
| Middle East & North Africa 1976–89; 1991–2001 | 14.72 | 4.70 | 0.80 | 25.54 | 0.12 | 0.06 | −7.09 |
| United Kingdom 1971–2001 | 3.70 | 5.21 | 0.32 | 1.20 | 0.00 | 0.00 | 7.38 |
| United States 1970–2001 | 5.73 | 5.62 | 0.42 | 1.95 | 0.05 | 0.00 | 8.94 |

*Source* Replication of Table 1 in Arrow et al. (2004)

**Table 5.9** Growth rates of per capita genuine wealth

| Country | Genuine investment as percent of GDP | Growth rate of unadjusted genuine wealth | Population growth rate | Growth rate of per capita genuine wealth—Before TFP adjustment | TFP growth rate | Growth rate of per capita genuine wealth—After TFP adjustment | Growth rate of per capita GDP |
|---|---|---|---|---|---|---|---|
| Bangladesh | 7.14 | 1.07 | 2.16 | −1.09 | 0.81 | 0.30 | 1.88 |
| India | 9.47 | 1.42 | 1.99 | −0.57 | 0.64 | 0.54 | 2.96 |
| Nepal | 13.31 | 2.00 | 2.24 | −0.24 | 0.51 | 0.63 | 1.86 |
| Pakistan | 8.75 | 1.31 | 2.66 | −1.35 | 1.13 | 0.59 | 2.21 |
| China | 22.72 | 3.41 | 1.35 | 2.06 | 3.64 | 8.33 | 7.77 |
| Sub-Saharan Africa | −2.09 | −0.31 | 2.74 | −3.05 | 0.28 | −2.58 | −0.01 |
| Middle East/North Africa | −7.09 | −1.06 | 2.37 | −3.43 | −0.23 | −3.82 | 0.74 |
| United Kingdom | 7.38 | 1.48 | 0.18 | 1.30 | 0.58 | 2.29 | 2.19 |
| United States | 8.94 | 1.79 | 1.07 | 0.72 | 0.02 | 0.75 | 1.99 |

*Source* Replication of Table 2 in Arrow et al. (2004)

An empirical application, a calculation of the sustainability of development in several countries and regions, is given by Arrow et al. in Tables 1 and 2, reproduced here in Tables 5.8 and 5.9.

# Chapter 6
# Climate: The Ultimate Resource?

**Abstract** The final chapter treats what might be considered a "new" or unconventional resource: global climate, which obviously affects the environment and also underlies the productivity of renewable resources and agriculture. The starting point is a brief discussion of a perceived disconnect between natural scientists and economists on the importance of climate change, with the former believing it to be perhaps the most important environmental problem of the century, requiring prompt and dramatic action and the latter seeing it in less compelling terms—though not all economists are in agreement here. Recent projections by the Intergovernmental Panel on Climate Change (IPCC) are given, with a discussion suggesting these are likely to be conservative in a variety of ways, including neglect of methane feedback, and for a variety of reasons, including the need for consensus among the parties. A discussion of potential impacts focuses on some often neglected and potentially catastrophic, due to tipping points and extreme events, which can in turn lead to major loss of capital (ports, buildings, coastal agriculture, and so on) not well captured in policy models. Given the time scales involved, discounting is of course key. As considered here, the issue boils down to the choice of how the pure (social) rate of time preference in the Ramsey equation should be specified. Two schools of thought are identified. One argues that it is an ethical choice, reflecting relative weights of different generations, and thus exogenous to the economic problem (though this is consistent, in the equation, with a positive discount rate even if the pure rate of time preference is zero). The other school argues that the rate of time preference must be consistent with observed rates of return on investment in private capital markets. Closely related to discounting is the topic of irreversibility, since the world will be locked into a changed climate and its consequences, such as rising sea levels, essentially forever on human time scales. Again there is a split between (some) economists and climate scientists, with the former pointing out that investment in new energy sources and facilities is also irreversible, and might dominate under certain circumstances. The discussion here focuses on a comparison of time scales associated with the two types of irreversibility. Finally, implications for policy are briefly discussed, including the advantages of a carbon tax, but also a potential problem: to achieve a desired objective the tax might have to be unrealistically high. In this case it might be supplemented by a negative tax in the form of a tax credit on energy conservation and renewables, though this can be distortionary (favors one energy source or technology



A. C. Fisher, *Lecture Notes on Resource and Environmental Economics*, The Economics of Non-Market Goods and Resources 16,
https://doi.org/10.1007/978-3-030-48958-8_6

over another). Some attention is also given to the role of public investment in basic research into innovative technologies that do not involve the emission of greenhouse gases.

## 6.1 Introduction

A question might be raised as to how a discussion of climate change fits in a volume largely devoted to the use of extractive resources, both nonrenewable and renewable, and prospects for exhaustion or overuse, There are a couple of links. First, there is a kind of analog to the dynamic problem of depletion: the deposition over time of carbon dioxide (CO2) and other greenhouse gases in the atmosphere due to the combustion of fossil fuels. Of course this involves an undesirable buildup, as opposed to an undesirable depletion (of a valuable resource), but some of the same analytical methods can be used in an optimization model (for costly control over time of greenhouse gas emissions). A second and much broader link comes from the recognition that climate underlies the availability of major renewable resources such as forests and fisheries as well as providing the conditions necessary to agricultural production. And then there is of course the environmental link. The alternative, non-extractive uses of forest resources such as tourism and outdoor recreation, and non-consumptive uses such as the survival of species in the wild, both plant and animal, are obviously dependent on climate and potentially affected by predicted movement to a warmer and generally dryer environment, with accompanying increases in extreme and catastrophic events such as droughts, floods and wildfires, each with consequences for resource use. The discussion that follows is drawn from a working paper (Fisher and Le 2014a), in addition to earlier lectures.

It seems fair to say that climate scientists, and natural scientists generally, believe that climate change is a major problem, perhaps the most important environmental problem facing us this century, already having at least local impacts and sure to have more over the coming decades, possibly catastrophic, as for example the unexpectedly rapid melting of Arctic ice, somewhat surprisingly including the Greenland ice sheet, leading toward the end of the century to a dramatic rise in sea level. A very recent discovery, that the formerly stable ice sheet in northeast Greenland is now melting rapidly, is particularly alarming (Khan et al. 2014), bringing nearer in time inundation of coastal areas and ocean acidification with devastating impacts on marine life and potential disruption of thermohaline circulation (IPCC 2013). A second discovery, independently by two teams of researchers along the same lines is that a precursor (a melting glacier) to early stage collapse of the West Antarctic Ice Sheet has begun and is irreversible (Joughin et al. 2014; Rignot et al. 2014). The potential for sea level rise and other adverse impacts of climate change was noted already in 2007 in the Fourth Assessment Report (AR4) of the Intergovernmental Panel on Climate Change (IPCC). However, there is in my judgment greater urgency since then as a result perhaps of the new discoveries, and of greater than expected increases in both melting and emissions of greenhouse gases (GHGs). The run of extreme weather events including heat waves, droughts, wildfires, and intense storms in the U.S. and elsewhere in recent years, all predicted consequences of climate change, is probably contributing to a growing popular perception as well that climate change is real, and is a problem to be taken seriously.

One could argue that the extreme events of the past few years represent simply fluctuations in a stable climate regime, but even if this were true, they are important as indicators of expected conditions under further warming. And this argument has become more difficult to make, as a result of a recent study by research teams for NOAA and the UK Met Office which detected the "fingerprints" of climate change on about half of the 12 most extreme weather events of 2012 (Peterson et al. 2013). For example, climate change helped raise the temperatures during the run of 100 F days in the American heat wave; drove the record loss of Arctic sea ice; and fueled the devastating storm surge of Superstorm Sandy—an interesting case, less obvious than heat waves and melting ice, but linked to climate change by the rise in sea level of nearly a foot at New York and along the New Jersey coast. Again, it's important to recognize that this storm is not just a one-time event, but an example of what by midcentury will be, according to a prominent geologist, "the new norm on the Eastern seaboard" (Wikipedia 2014).

On the contrary, the impression one gets from much of the established economic literature is that global warming is not an issue that requires dramatic action in the near term. The story goes that a prominent economist who, on being asked what he thought ought to be done about the recently identified problem of climate change, responded, "What's the problem? When I cross from Washington, DC to Virginia and it's five degrees warmer, I take off my jacket." But even serious analyses suggest that only modest steps (relative to those recommended by climate scientists) to control GHGs are appropriate at this time, with a gradual increase in stringency toward the end of the century, the usual time horizon for simulations in integrated assessment models (IAMs) such as DICE (Nordhaus 1993, 2007a), FUND (Tol 1997), and PAGE (Hope 2006). This is consistent with the apparent empirical findings of minor or potentially even positive impacts of warming on U.S. agriculture, considered the most vulnerable sector of the economy (Mendelsohn et al. 1994; Deschenes and Greenstone 2007).

To be fair, economists have not spoken with one voice. With respect to agriculture, other empirical studies suggest a rather different outcome: substantial and significant negative impacts toward the end of the century under a "business as usual" scenario (Schlenker et al. 2005, 2006; Fisher et al. 2012) and attribute the earlier and more optimistic findings to conceptual and statistical errors. Though there are conflicting results for the U.S., there seems to be general agreement on very substantial negative impacts on many developing countries, perhaps catastrophic for those already facing food insecurity (Mendelsohn and Dinar 1999; Cline 2007). Moreover, it's my sense, from the vantage point of 2017, that the outlook of economists is evolving, due perhaps to the emergence of a new generation of climate economists who are producing detailed and careful empirical analyses of the impact on both the environment and the economy, on individual energy sectors and in the aggregate, of alternative policies such as carbon taxes and cap-and-trade schemes. Some of this work is cited in the sections that follow.

And of course on a macro level the *Stern Review* 2007, using the PAGE model, gives serious attention to potential damages from unchecked warming, and argues that strong action should be taken in the near term to reduce emissions of GHGs. The *Review*, commissioned by the British government, has received a great deal of attention both within the economics community and beyond. The reception by economists has however generally not been favorable. Reviews tend to be critical, arguing that Stern's policy recommendations are unwarranted, stemming from some combination of an over-estimate of potential damages, and an unrealistically low discount rate and/or elasticity of the marginal utility of consumption assumed in the model simulations. Widely cited critical reviews include those by Nordhaus (2007b) and Weitzman (2007) in the *Journal of Economic Literature*, although Weitzman suggests that Stern may in fact be right in his recommendations but for the wrong reasons, not the right one, the Weitzman hypothesis on the importance of potentially catastrophic outcomes in the "fat tails" of a distribution (Weitzman 2009). A *Symposium on Climate Change* in the *Review of Environmental Economics and Policy* a year later includes critical reviews by Mendelsohn (2008) and Weyant (2008), along with one by Sterner and Persson (2008) that takes, as the paper's title suggests, an "even sterner" view of the need for urgent action.

So while the scientific community appears to take a near-unanimous view of the magnitude of the problem of climate change and the urgency of dealing with it, the impression one gets from much of the economic literature is mixed, with some of the more prominent empirical contributions suggesting that impacts will be modest (if not positive) in the most climate-sensitive sector of the U.S. economy, and IAM simulations that call for only modest steps to reduce emissions of GHGs in the near to medium term. My view is that much of the analysis to date has a tendency to understate the magnitude of the problem and perhaps also to overstate the costs of dealing with it. This is evident in a recent statement of the case against stringent controls, "The Climate Policy Dilemma", by Pindyck (2013b). In what follows I offer some reflections on climate policy stimulated by the paper, broadening the focus as we go to the literature more generally, both scientific and economic. Interestingly, in another paper published at about the same time, Pindyck (2013a) makes a case for what we might call an active control policy, if not necessarily immediately stringent, based on some of the same considerations that I shall argue form the basis for a relatively stringent policy.

In my view the main implication of "The Climate Policy Dilemma", skepticism regarding stringent controls on GHG emissions, though appropriately hedged due to the massive uncertainty surrounding future climate and impact variables, does not necessarily follow. I suggest some reasons for supporting a policy of relatively stringent controls. The difference in policy recommendations is due primarily to what I regard as questionable assumptions, both explicit and implicit, here and elsewhere in the literature, about future climate change, impacts, and costs of averting climate and other potential catastrophes. In part this is just a matter of updating some of the projections and findings based on recent research. In part it has to do with what I regard as important omitted impacts, and a different take on the role of discounting and irreversibilities, key issues in the context of climate change.

The next section reviews recent climate change projections, with attention to aspects often not included, and therefore not reflected in the IAMs or other policy analyses. Section 6.3 looks at potential impacts, including those not immediately translated to conventional measures of economic output or income and so not taken into account in the models. Section 6.4 offers some remarks on the sensitive issue of discounting in climate policy. Section 6.5 discusses relevant (and irrelevant) irreversibilities. Section 6.6 considers catastrophic climate change in the context of other potential catastrophes, and Sect. 6.7 concludes with a discussion of implications for climate policy.

## 6.2 Climate Change Projections

We begin with the most recent IPCC projection (IPCC 2013) of global mean temperature (GMT) in the highest concentration pathway (RCP8.5) for GHGs in the atmosphere, what we might call the "business as usual" case, characterized by continued heavy fossil fuel use not significantly constrained by climate policy: an increase of 3.7 °C, with a 90% probability range of 2.6–4.8 °C relative to 1986–2005 by the end of the century. Since GMT has already (as of 2014, the date of the lecture on which this section is based) increased by approximately 0.8 °C relative to the pre-industrial level, the 3.7 °C would look like 4.5 °C relative to the pre-industrial level, which is how the projected increase has typically been presented in the past. Either way, as I shall indicate, this falls well above the increase of 2 °C climate scientists have suggested must not be exceeded if we are to avoid a broad range of disastrous consequences.

There are reasons for believing that the IPCC estimate is in fact too low. First, it does not include the additional increase due to release from Arctic permafrost of $CO_2$ and methane, a gas with a much greater warming potential, though shorter atmospheric residence time. This may be defensible on the grounds that the cumulative release is uncertain at this time. However, it's already happening on a small scale, not sufficient to affect GMT, but certain to greatly increase under the temperatures projected by the IPCC, bearing in mind that temperatures at Arctic latitudes are increasing at least twice the rate of increases in the GMT, with a projected increase of approximately 8 °C or more by the end of the century. A sense of the regional variation in projected temperature increases is given in Fig. 6.1, based on an earlier IPCC report. The concern is that this is a positive feedback loop: the higher the temperature, the greater the release of methane and $CO_2$; the greater the release, the greater the increase in temperature; and so on.

Moreover, such increases in releases from the Arctic permafrost are not merely hypothetical. A recent study of the East Siberian Arctic Shelf (ESAS) conservatively estimates that the current methane release from submarine permafrost is 17 million metric tons annually, more than twice the level previously estimated (Shakhova et al.

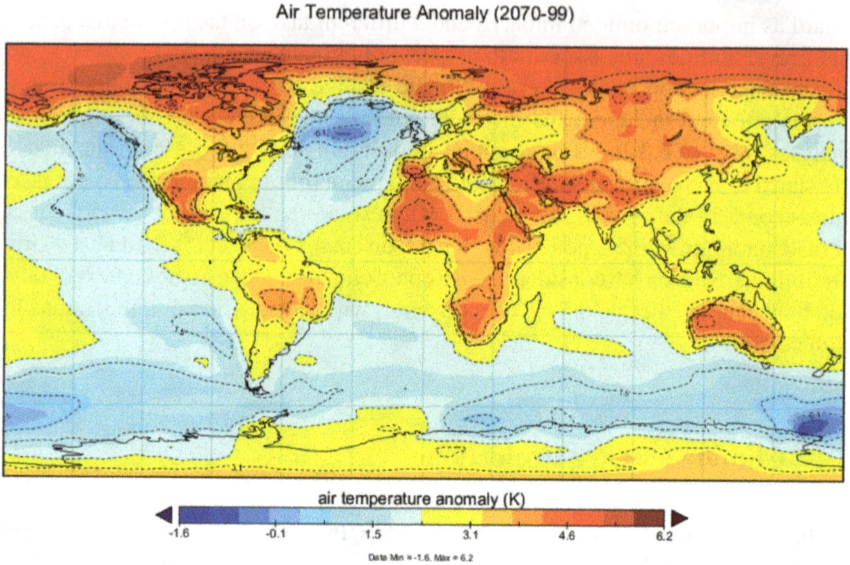


**Fig. 6.1** Projected temperature increases by the end of the 21st century under the SRES A2 emissions scenario (relatively high emissions, though not the highest). The temperature anomaly is averaged over the 2070–2099 period. The baseline is 1961–1990. *Source* IPCC AR4-2007

2013). This sounds like a lot, but it represents only 3% of total methane emissions of 500 million tons from all sources, natural and anthropogenic. This translates to 12.75 million tons of carbon, in turn a small fraction of the more than nine billion tons from other GHGs, mostly $CO_2$. There is still reason for concern, since the permafrost in the shallow waters off the ESAS has warmed closer to the thawing point than terrestrial permafrost, and increased storm activity may accelerate significant increases in methane emissions (Shakhova et al. 2013). In a slow-release scenario, emissions are projected to increase by 5% per year (Shakhova et al. 2010), resulting in a release of almost 200 million tons in year 50, and a cumulative total of more than 4 billion tons over the period. In fast-release scenarios, emissions jump dramatically, to the point where they are comparable to those of $CO_2$.

A warming ocean is likely to result in the breakdown of a substantial part of the huge reservoir of methane in undersea ice structures called clathrates. Shakhova et al. (2010) estimate that these contain as much as 1,750 billion tons of methane off the ESAS. Conservatively, they project breakdown and release of just 3.5%, or 50 billion tons, but in some scenarios over as short a period as five years, implying an average release of 10 billion tons per year, or 7.5 billion tons of carbon.

For the much larger global ocean clathrate reservoir, Archer and Buffett (2005) estimate a global inventory of about 5,000 billion tons of carbon in the stored methane, and eventual releases of 2,000–4,000 billion tons in response to a cumulative 2,000 billion ton anthropogenic atmospheric carbon release. These are very large numbers, but assuming annual emissions of GHGs, currently over 9 billion tons of carbon

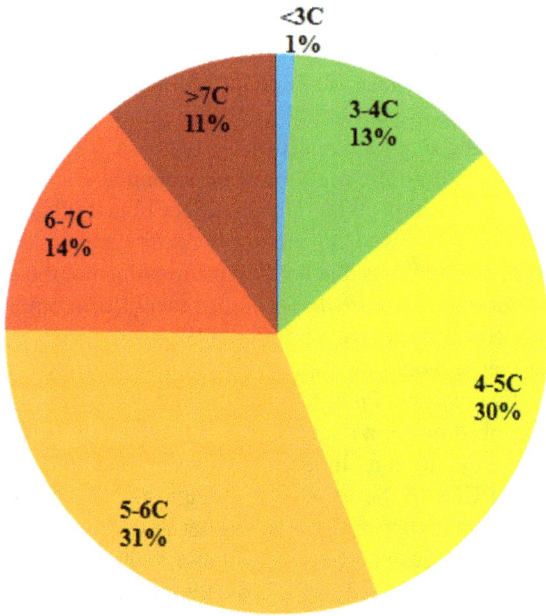


**Fig. 6.2** Probabilistic prediction of global average surface temperature under no policy by 2100 relative to 1990. The median value is 5.2 °C. *Source* Sokolov et al. (2009)

annually, of 10 billion tons, these targets would be reached within the next 200 years, sooner if emissions continue to increase.

A second reason for thinking the IPCC projections may be on the low side is that the IPCC tends to be conservative—recall the underestimate of GHG emissions and Arctic ice melting in AR4, in 2007. A possible explanation is that the report, in particular the *Summary for Policy Makers*, must be approved by all member governments, including those of large producers of fossil fuels, and in particular those whose export earnings and a substantial fraction of GDP derive from this source. This is not to suggest any attempt to misrepresent findings by the IPCC, but could in part explain the tendency to be conservative in drawing conclusions.

An alternative set of projections, from the MIT Integrated Global System Model (Sokolov et al. 2009) shows a median increase in GMT of 5.2 °C (again, + 0.8 °C for the increase from the pre-industrial level), with a range of 3.5–7.4 °C (+ 0.8 °C). Moreover, as Fig. 6.2 indicates, the distribution is not symmetric. The probability of warming more than 5 °C is 56%, less than 5 °C 44%. As economists we are perhaps not the best judges of the merits of alternative climate models, but the MIT model, given its comprehensive and thorough testing and documentation, with no apparent potential for bias, is certainly worth looking at alongside the IPCC projections. It appears that the MIT model projections, like those of the IPCC, do not include the permafrost feedback, so they would also need to be augmented.

Finally, there is the issue of the appropriate time horizon for projections and planning. Early discussions of climate change focused on the end-point of a doubling of the atmospheric concentration of $CO_2$ from the pre-industrial level of 280 ppm, rather than a particular date. More recently, in the IPCC reports AR4 and AR5, and in the models and related policy analyses, the end-point is the year 2100. Given current (again, as of about 2014) concentrations of 400 ppm for $CO_2$ alone and 470–480 for $CO_2$-equivalent (adding in other gases such as methane and nitrous oxide and converting to an amount of $CO_2$ required for equivalent warming), and assuming a continuation of current rates of accumulation, the doubling will be reached well before the middle of this century. Since climate change is clearly a long-run problem, the shift in focus to the end of the century makes sense—as far as it goes.

The difficulty is that neither the doubling nor the end of the century represents an equilibrium, an end to the warming process. In a comprehensive, rigorous and remarkably foresighted analysis, Cline (1992) calculates that an appropriate end-point in this sense would be approximately the end of the 23rd century, or the year 2300, for two reasons. First, fossil fuel (mainly coal) reserves would be either exhausted or choked off by a rising price due to scarcity (note, even assuming no restrictions on use due to non-climate-related environmental impacts). Second, mixing of dissolved $CO_2$ in the upper ocean with the deep ocean becomes important in about 200 years, exposing a much greater reservoir of water to mixing and thus greatly increasing ocean absorption. A conservative estimate of the increase in GMT by this date is approximately 10 °C, with a plausible range including substantially greater increases (Cline 1992, pp. 56–58) and of course greater still in the Arctic, with the implications this entails for melting ice and methane and $CO_2$ release.

Apart from another early discussion of the end-point issue, in Fisher and Hanemann (1993), there seems to be little if anything in the subsequent economic literature, perhaps due to the focus, first on a doubling of atmospheric concentrations of GHGs and then on the end of the 21st century in the scientific literature. Recently however some estimates of what might be expected by the year 2300 have appeared, consistent with the early calculations by Cline. The German Climate Computing Center (DKRZ 2014), in simulations conducted for AR5, projects an increase of nearly 10 °C from the 1986–2005 level in an extension of the RCP8.5 scenario to 2300, as shown in Fig. 6.3.

Other recent long-term projections reported in the *New Scientist* similarly suggest that a "burn everything" scenario could lead to atmospheric concentrations of $CO_2$ as high as 2000 ppm, in turn leading to a global temperature rise of 10 °C (Marshall 2011). The report includes still more alarming speculations about the danger of runaway processes, once temperatures reach this level. And finally, a study in the Proceedings of the National Academy of Sciences (PNAS) projects possible eventual warming of 12 °C from fossil fuel burning. The main point of the study is however the novel and sobering one that at these temperatures regions currently holding the majority of the human population would become uninhabitable due to induced hyperthermia as dissipation of metabolic heat becomes impossible (Sherwood and Huber 2010). This brings us to potential impacts of projected climate changes, the subject of the next section.

I should make clear that these are "worst case" scenarios, implied by unconstrained emissions of GHGs, to the year 2100 and then beyond, to the year 2300. I am not predicting that this will happen, for several reasons including innovations in non-

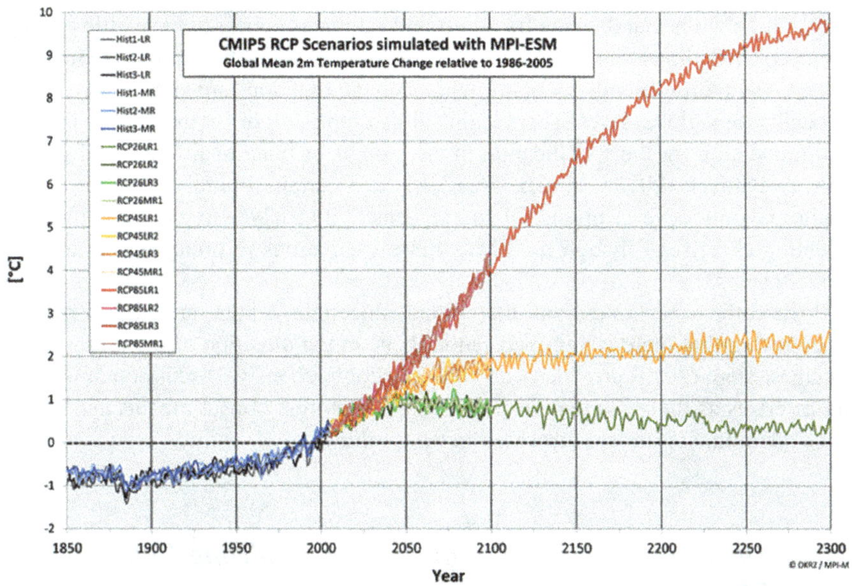


**Fig. 6.3** Projected global mean temperature (GMT) by 2300. *Source* DKRZ (2014)

fossil energy technologies, in energy efficiencies, and in carbon capture and storage (CCS) that may make possible the continued burning of the very large reserves of coal (though with other health and environmental costs, as noted later on). The point is rather that, without a clear understanding on the part of economists and decision-makers of the potentially catastrophic impacts of unconstrained emissions, it becomes difficult to see how the analyses and policies that would lead to a timely development of these alternatives will occur.

## 6.3 Potential Impacts and Problems of Estimation

Considering first the impacts that go into the economic models used in policy analyses, there are major difficulties with the estimates of potential damages in the IAMs, and therefore the optimal control trajectory for GHGs in the models. This means that the social cost of carbon (SCC), calculated from model outputs and the basis for policy analyses and rule-making by governmental agencies, also comes into question. The models represent an important step in the multidisciplinary analysis needed to address climate change, and illustrate the links between climate and economy, given assumptions about a variety of functional forms and parameters. The difficulty comes in the use made of the numerical results of model simulations to make inferences about policy.

One problem is that damage functions and estimates appear to have little relation to empirical findings based on studies of sectoral impacts, both physical and economic. And more general, economy-wide damage functions are simply not known, especially on a global level. There is thus little empirical, or for that matter theoretical, foundation for the specification in the models of functional forms and parameters, as Pindyck (2013a) argues. This leads to a certain arbitrary character of the results and policy prescriptions. Of course, sensitivity analysis as carried out in most modeling exercises—though not always in presentations to policymakers and the public—helps.

Beyond the arbitrariness and the obvious uncertainty, there are several reasons for believing that there is very substantial bias, in the direction of underestimating damages. Stern (2013) provides a detailed and comprehensive discussion of what he characterizes as the gross underestimations in economic impact models and IAMs. Some additional issues are discussed in what follows.

### *6.3.1 The Importance of Non-linearity in Modeling Climate Change Impacts*

One type of nonlinearity, generally not captured in the models, is represented by catastrophic events and the possibility of tipping points in climate systems—the threshold beyond which climate change can no longer be accommodated. One example of this is, as noted in the last section, under a 12 °C increase in GMT, projected for the end of the 23rd century with no constraints on emissions of GHGs, regions currently holding the majority of the human population would become uninhabitable due to induced hyperthermia as dissipation of metabolic heat becomes impossible (Sherwood and Huber 2010).

Another tipping point can come from sea level rise. The increase of 4–5 °C in GMT projected for the end of the 21st century in a business-as-usual scenario would lead to permanent abandonment of living spaces such as islands and low-lying areas, including large portions of some countries, inundated by the resulting rise in sea levels of at least one meter (World Bank 2012). Adding in the effect of (complete) melting of the West Antarctic Ice Sheet (WAIS), a serious concern for the period beyond 2100, with some possibility of a more rapid rise, the projected rise in sea level would be much greater. Earlier estimates are in the range of 5–6 m, with a more recent calculation based on improved bedrock and surface topography setting it at 5 m, but as low as 3.3 m taking into account the proportion of the bedrock above sea level, with the ice sheet overlay not subject to rapid disintegration (Bamber et al. 2009). On the other hand, there would be regional variation in the impact, with the maximum increase concentrated along the Pacific and Atlantic seaboard of the U.S, where the value would be about 25% greater than the global mean. The Greenland Ice sheet, already melting, would ultimately add about 6.5 m (Poore et al. 2011). Reduction of both similar to past reductions in the geologic record would add

**Table 6.1** Examples of potentially catastrophic impacts

| | |
|---|---|
| 1. Increase of 4–5 °C in GMT projected for the end of the 21st century would lead to permanent abandonment of living spaces such as islands and low-lying areas, including large portions of some countries, inundated by the resulting one meter rise in sea levels (World Bank 2012) | 2. Melting of the West Antarctic Ice Sheet, a serious concern for the period beyond 2100: the projected rise in sea level would be much greater. Recent findings suggest that this has already begun and is in its very early stages (Joughin et al. 2014; Rignot et al. 2014), with complete melting resulting in anywhere from 3.3–5 m global average increase, depending on assumptions, and about 25% higher along the east and west coasts of the U.S. (Bamber et al. 2009) |
| 3. The Greenland Ice sheet, already melting, would ultimately add about 6.5 m (Poore et al. 2011) | 4. Reduction of both similar to past reductions in the geologic record (presumably not complete) would add at least 10 m, resulting in flooding of about 25% of the U.S. population, with the major impact on people and infrastructure along the East and Gulf Coast states (Poore et al. 2011) |
| 5. During the Pliocene Epoch, 2.4–5 million years ago, which was about 3 °C warmer than today, sea levels were as much as 15–25 m higher, suggesting that part of the much larger East Antarctic Ice Sheet, generally believed to be more stable, must have been eventually vulnerable to melting (Hansen et al. 2013) | 6. An increase of 12 °C in GMT, projected for the end of the 23rd century with no constraints on emissions of GHGs: regions currently holding the majority of the human population would become uninhabitable due to induced hyperthermia as dissipation of metabolic heat becomes impossible (Sherwood and Huber 2010) |

at least 10 m, resulting in flooding of about 25% of the U.S. population, with the major impact on people and infrastructure along the East and Gulf Coast states. As with potential methane releases, it could get a lot worse. During the early Pliocene, which was about 3 °C warmer than today, sea levels were as much as 15–25 m higher, suggesting that part of the much larger East Antarctic Ice Sheet, generally believed to be more stable, must have been eventually vulnerable to melting (Hansen et al. 2013). Potential impacts from extreme heat and sea level rise are displayed in Table 6.1 and Fig. 6.4a and b.

In response to criticism that damage functions that fail to capture nonlinearities caused by extreme events or abrupt changes understate potential damages, Nordhaus and Sztorc (2013) add 25% to the DICE damage function to account for all non-monetized impacts. This seems arbitrary at best, and almost certainly still an underestimate, especially as we look beyond the end of the century. For example, Ackerman et al. (2010) show, using the standard DICE model, that temperature increases of up to 19 °C can involve a loss in output of 50%. While this seems like a lot, in light of the globally catastrophic impact of a 10 °C increase it seems doubtful that human civilization could survive the much larger increase. But tinkering with the functional form of the (usually specified as quadratic) damage function to avoid this

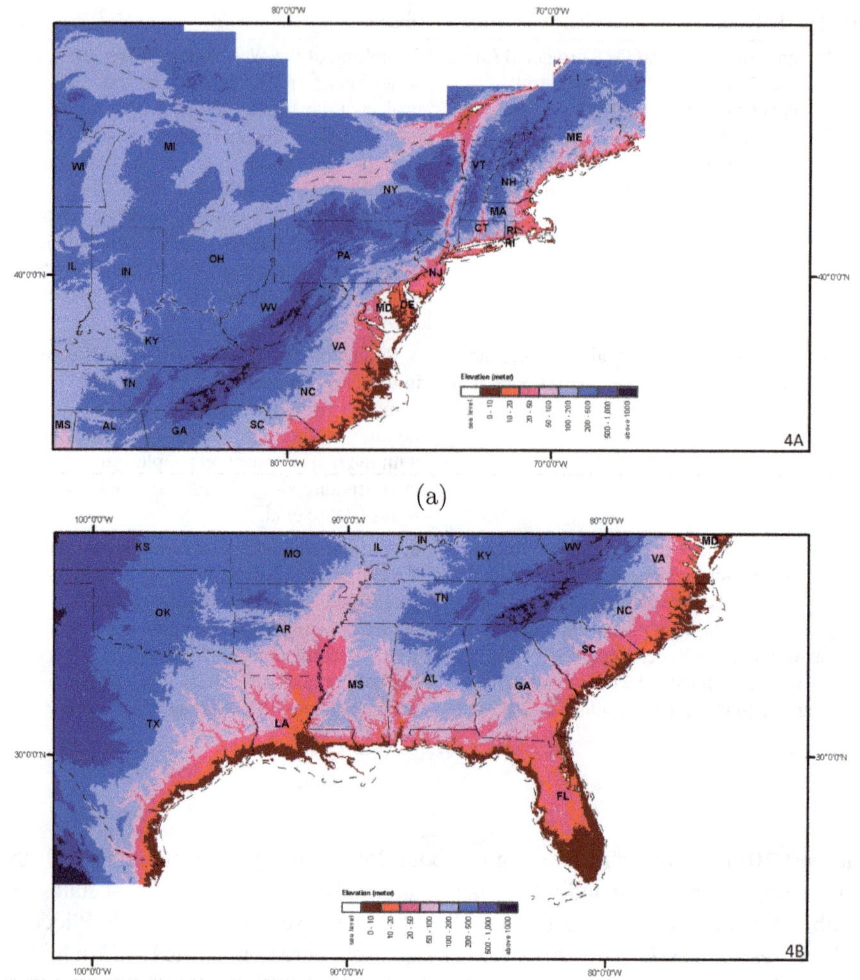


(a)

(b)

**Fig. 6.4** Risk of sea level rise (inundation; transient flooding already occurring), North East and South East/Gulf Coast states. *Source* USGS 2.5-arcmin Digital Elevation Model

sort of unrealistic outcome, as some have suggested, also seems arbitrary, a serious concern in light of the sensitivity of results to functional specifications. Raising the temperature exponent from 2 to 3 in a standard damage function causes damages to jump from 10.4% of output to 33.3% when the elasticity of marginal utility is 1, or from 3.3% to 29.1% when the elasticity is 2 (Stern 2008).

A more promising approach to modifying what are in my judgment clearly unsatisfactory damage functions is taken in recent (once again, as of about 2014) papers that introduce explicitly the possibility of tipping points and extreme or catastrophic events, along with modeling of the associated uncertainty. Ceronsky et al. (2011) use the FUND model to estimate the SCC under what they characterize as low-probability, high-damage events, including large scale release of oceanic methane hydrates (clathrates) and climate sensitivity (response of temperature to a given increase in atmospheric concentrations of GHGs) above "best guess" levels. The mean SCC in probabilistic simulations is as much as three times conventional estimates—and much higher still in the most extreme scenarios coupled with a low or zero pure rate of time preference. The conclusion is that the potential for catastrophic outcomes of the sort modeled here can justify aggressive near-term mitigation, in contrast to the comparatively modest steps called for when this is neglected. In another example, Lemoine and Traeger (2012) study the impact on the optimal carbon tax in a modified DICE model of two kinds of tipping points, one involving climate sensitivity and the other a longer carbon cycle. The optimal near-term carbon tax is increased by up to 45%, and peak warming reduced by 0.5 °C compared to a model without possible tipping points.

An approach along very different lines, empirical damage functions based on the observed response of human systems to climate variability, is suggested by Kopp et al. (2013). This is supported by some work on potential damages in the most extensively studied sector of the U.S. economy: agriculture. In the series of econometric studies cited earlier by Schlenker, Hanemann and Fisher, and a separate study by Schlenker and Roberts, a nonlinear impact provides a better foundation for increased damages than an arbitrary add-on or change in exponents. The nonlinearity is exhibited in two ways. First, a dramatic increase over the next several decades in extreme heating days, measured by growing degree days at temperatures above 34 °C, is responsible for nearly all of the potential negative impact of warming on farmland value in the U.S. estimated in Schlenker et al. (2006). Projected increases in a measure of average temperature increase are much less important. Second, crop yield functions exhibit sharply diminishing returns after an optimal temperature has been reached (Schlenker and Roberts 2009). Yields are stable or slowly increase with temperature up to 29 °C for corn and 30 °C for soybeans, and then fall sharply with each additional degree, as shown in Fig. 6.5.

An interesting question is, to what extent might this sort of relationship hold in other areas or activities affected by climate change, including ecosystem functioning, and might it be appropriate for modeling damages in IAMs? It seems plausible, but more empirical research will be needed to confirm this.

### *6.3.2 Impacts on the Environment, Extreme Events, and Capital Losses*

Direct impacts of climate change on goods and services provided by the natural environment are in effect omitted in much of the literature on damage estimates, including those in the IAMs, due to the implicit assumption of perfect substitutability

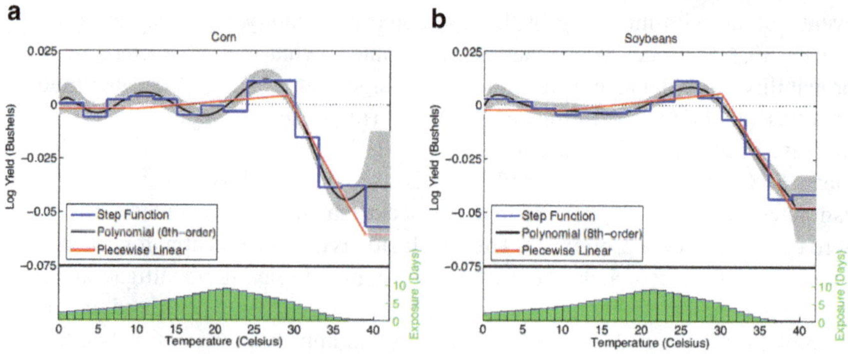


**Fig. 6.5** Nonlinear impact of temperature on crop yields. *Source* Schlenker and Roberts (2009)

between environmental and market goods. This means, as noted by Sterner and Persson (2008), that "a dollar's worth of climate damages, regardless of the kind, can be compensated by a dollar's worth of material consumption." They go on to argue that "if there are limits to substitutability... then our analysis of climate change needs to take into account the content of future growth." In particular, it is the environmental goods that can be expected to become relatively scarce, so that their shadow prices will be rising (for an early analysis along these lines, though not in relation to climate change, see Fisher and Krutilla (1975)). As they illustrate using the DICE model, making some plausible (though admittedly speculative) assumptions about the value of nonmarket environmental goods and their (limited) substitutability by market goods, the implied change in relative price has an effect on optimal abatement very similar to that produced by adopting the near-zero pure rate of time preference in the *Stern Review*, while still keeping the higher rate used in DICE and preferred by some of the critics. Making both changes would call for still more dramatic reductions in GHG emissions over the next several decades.

As environmental impacts such as heat waves, wildfires, droughts, and intense storms become more frequent and more severe, they will undoubtedly affect consumption included in the measured national income accounts as well, especially where they result in destruction of coastal infrastructure—which as we note below is not picked up in IAM damage functions—and especially in developing countries. A relevant and somewhat surprising finding here is reported in the recent paper by Hsiang and Jina (2013) on the long-run impact on economic growth of a major tropical storm. They find on the basis of exhaustive econometric analysis of several datasets that national incomes decline relative to their pre-disaster trend and do not recover within 20 years, and that the cumulative effect of this persistently suppressed growth is both significant and large: a 90th percentile event reduces per capita incomes by 7.4% two decades later.

One explanation may be that the immediate impact of the event is to destroy infrastructure, which in turn impacts productivity in subsequent periods. This raises a more general point—first noted by Fisher and Hanemann (1993) and discussed at some length in Stern (2013)—about estimates of losses associated with climate change, in particular in the IAMs. In their focus on losses in income or output, these models don't capture capital losses, which in my judgment are likely to be more important in assessing extreme events such as the recent Superstorm Sandy and other major tropical storms, with their extensive damage to coastal infrastructure and subsequent impact on productivity.

Finally, a couple of recent studies that similarly expand our focus beyond conventional damage estimates suggest that losses may be more widely experienced than previously believed. Heal and Park (2013), drawing on a variety of data sources, both physiological and economic, demonstrate a link between temperature and income, identifying a link between temperature and labor productivity as a likely cause, with a broad measure of productivity falling beyond temperatures in the range of 18–22 °C. And although not directly linked to GDP or productivity, a relationship between temperature and conflict, ranging from domestic violence to regional conflicts has been found by Hsiang et al. (2013). The message from these studies, including Hsiang and Jina, is that potential impacts of climate change over the next several decades are likely to be wide-ranging, going well beyond those included in earlier sectoral studies or conventional damage estimates.

## 6.4 Discounting

Any discussion of climate policy inevitably turns to the question of how to discount the potential damages associated with climate change, since damages are expected to become increasingly severe over the next several decades and beyond. This time frame, along with the magnitude of the problem, makes the discounting decision more critical than in the typical investment problem. Not surprisingly, optimal policy in the DICE model is more sensitive to changes in the discount rate (keeping within a plausible range) than to changes in other variables in alternative runs of the model (Nordhaus 1993, 2007a)—although as noted earlier, taking account of changing relative values over time of environmental and market goods, with limited substitutability of the latter for the former, can have a similarly dramatic effect.

Since the discussion on discounting in the climate problem is often set in the framework of the Ramsey equation, we write it below.

$$r = \rho + \theta * g$$

In words, the consumption discount rate $r$ is equal to the sum of two components: the pure rate of time preference $\rho$, and the product of the elasticity of the marginal utility of consumption $\theta$ and the rate of growth in per capita consumption $g$. Controversy centers on the choice of $\rho$, and to a lesser extent on $\theta$. Of course there can also be differing expectations concerning $g$—a point we'll come back to, as in our judgment it becomes important in thinking about climate policy. This equation can be adjusted to take into account uncertainty, as in Heal (2009) and Arrow et al. (2014), but as the adjustment does not affect the discussion below, we need not develop it here.

Pindyck (2013b) presents a hypothetical example involving two individuals, John and Jane, deciding how much to spend and how much to save of their respective (equal) incomes, with John saving 10% each year to help finance the college education of his perhaps yet unborn grandchildren and Jane saving nothing, rather spending on sports cars, boats and expensive wines. He then poses the question of whether John's concern for his grandchildren makes him more ethical than Jane, and answers it by saying that economists don't have much to say about the question. We suspect the average person, confronted with these choices, might not hesitate to label John's choice the more ethical one. Or maybe not. In any event we agree that, as economists, we have little to say concerning John's or Jane's individual choice.

Does this example shed any light on the choice of the rate of time preference in the climate change problem? Many, and we would guess most, people, even most economists, would say that in the very different setting of a social choice: how much, if anything, the present generation should do to ward off major and perhaps catastrophic impacts of climate change, some concern for the welfare of future generations—including Jane's yet unborn grandchildren—is the ethical choice. In the context of the Ramsey equation, what this means is that $\rho$ should be understood as the pure *social* rate of time preference, not the private rate implicit in the example.

This leads to the question of how the rate of time preference should be specified. Although there is a vast literature on the question, even in the context of climate policy, at the risk of oversimplifying we can distinguish two approaches. One, implicit in Pindyck's example, used throughout the *Stern Review*, and advocated in a wide-ranging review of climate economics by Heal (2009), considers that the choice is an ethical one, "a decision on the relative weights of different generations of human beings." Unlike the consumption growth rate, and to some extent the elasticity of marginal utility, the social rate of time preference is, in Heal's words, "exogenous to the economic problem." Once this is set, the discount rate, $r$ in the Ramsey equation, can be determined by adding in the $\theta * g$ term. It's important to note that this will ordinarily imply a positive discount rate even if the pure rate of time preference is zero, that is, there is no preference for the present over future generations. If, for example, the elasticity $\theta = 1.5$ and the consumption growth rate $g = 2\%$, the discount rate $r = 3\%$.

Returning to the question of the consumption growth rate, note that adjusting this to reflect "consumption" of the non-market services of the natural environment that support the human economy, and that are expected to be increasingly adversely affected by climate change, this rate, and therefore the discount rate in the Ramsey equation, will be reduced, and could conceivably become negative at some point, especially if accompanied by a low or zero rate of time preference. Alternatively, following Heal, we can distinguish two separate discount rates, one corresponding to ordinary consumption goods and the other to nonmarket services of the environment, with the former clearly positive and the latter possibly negative. Negative discounting, where appropriate, would of course mean that future losses look larger, not smaller, from the perspective of the present.

The other approach to the specification of the pure rate of time preference is exemplified in the DICE model (see Nordhaus (2007a)), where the left-hand side, the discount rate $r$, is explicitly chosen to be consistent with observed rates of return on investment in private capital markets. From this, along with the second term on the right-hand side, $\theta * g$, there is an implied pure rate of time preference (positive), thus bypassing any discussion of what constitutes an ethical choice by the current generation. There are two difficulties, in my judgment, with this approach. First, it implicitly assumes that the private rate of time preference, not the social rate, is relevant to a social choice on climate policy. And second, it measures $g$ solely by the growth rate in per capita consumption of conventional market goods. In fairness, the same criticism (neglect of impacts on nonmarket goods) could be made of most discussions of climate policy that take the alternative, ethical choice approach.

The two approaches are discussed at some length with special reference to the climate problem by Heal (ethical) and Nordhaus (market), proving perhaps that this is a subject on which reasonable economists can differ. That said, my view is that there is no necessary connection between interest rates or rates of return in private capital markets and the pure rate of (social) time preference appropriate to climate policy decisions.

## 6.5 Irreversibilities

Pindyck (2013b) raises an interesting point in the context of the discussion on discounting, though not limited to discounting. He argues that "the case for a stringent climate policy should be reasonably robust and not rely heavily on the value of a particular parameter (in this case the rate of time preference)." But it seems to me that one might equally well argue that the case for a modest (as opposed to stringent) climate policy, one characterized by modest controls in the near to medium term, should instead be reasonably robust. We are confronted with the possibility of two types of errors: Type 1, that a very modest policy will lead to disastrous climate consequences; and Type 2, that a stringent policy will lead to unnecessarily large mitigation and adaptation expenses. While neither outcome is desirable, the former strikes me as more important to avoid in that it can impose extraordinary costs for centuries, or millennia, depending on the nature of the impacts, whereas the latter is reversible in a few years or at most a few decades—and at relatively little cost if done within the normal replacement cycle of capital.

In this connection we note that one of the main sources of concern about climate change in the scientific community is precisely that in important respects it is irreversible. Emissions are reversible only over a very long period that stretches well beyond the useful life of a piece of capital equipment or the time horizon in economic planning models; once in the atmosphere they persist for many decades or even centuries. Further, impacts such as the inundation of large areas and loss of species, among many others, are essentially irreversible on human time scales.

Paradoxically, in a number of studies economists have argued that the relevant dynamics in fact suggest less control, not more. The control decision is modeled as an investment, for example in a new, non-fossil-fuel-based energy system, and the investment is also seen as irreversible. Simulation modeling based on what look like plausible assumptions from the DICE model about climate and economic parameters suggests that the investment irreversibility dominates. The decision is further tilted toward holding off on control of emissions taking into account uncertainty about the degree of climate change and the impacts. If learning is possible, the implication is to refrain from some or all of the investment in reducing emissions today in order to make a better decision tomorrow on the basis of better information about the potential benefits (see for example Kolstad (1996); Fisher and Narain (2003); Pindyck (2007)). The difficulty here, in addition to the underestimate of long run damages from climate change, is that the models fail to capture a key feature of the decision problem: *the time scales for climate and investment irreversibilities are not the same—in fact are likely to differ by orders of magnitude. Treating them symmetrically will produce a misleading result.*

Another problematic feature of the models is the treatment of the decision to invest in a new energy system as once-and-for-all, all-or-nothing. But the decision, or in practice the decisions, are not one-time, all-or-nothing. At any decision point, say one of the 10-year time steps in a model such as DICE, in reality an array of choices presents itself, and the outcome will tilt the mix of fossil fuel, renewable source, and energy-efficiency, capital in one direction or another—until the next decision point. Further, once we disaggregate "investment" in this fashion, a relevant symmetry becomes apparent. Investment in, say, a fuel-efficient car, or an array of solar collectors, or a facility to generate electricity from biofuels, might be considered irreversible in the short to medium term. But so is investment in another coal-burning plant to generate electricity, or another SUV getting 11 miles to a gallon of gasoline.

The conclusion we draw from this discussion is that *over the uniquely long period relevant to discussion of climate policy, the irreversibility— and therefore the option value—that matters is with respect to climate change and its impacts, and not investment in one or another energy facility.*

## 6.6 Catastrophic Climate Change in Context

Before considering climate change in the context of a world in which other potential catastrophes loom, we should indicate what might constitute catastrophic climate change. The scientific consensus is that stabilization of atmospheric $CO_2$ concentrations at 450 ppm (which implies a $CO_2$ equivalent, $CO_2$eq, level of about 550 ppm adding in likely concentrations of other GHGs such as methane), in turn roughly consistent with an increase in GMT of 2 °C toward the end of the century relative to the pre-industrial level, will be necessary to keep climate impacts from becoming likely or very likely to reach what most, and certainly most scientists, would regard as catastrophic levels in terms of their impact on the natural systems that support human economic activity and even on the economy directly.

Pindyck (2013b) frames the discussion of how potential climate catastrophes are to be taken into account in climate policy, in view of the potential for catastrophes in other areas, in terms of the willingness to pay (WTP) to avert or reduce the probability of occurrence of each type of catastrophe (climate, mega-virus, detonation of one or more nuclear weapons in a major city, and so on), taken alone. Although WTP to avert climate catastrophes may be as high as 10% of GDP (Pindyck's illustration), once we recognize that there may be a similar WTP for the others as well, we are confronted with something like 60–70% of GDP to avert all of the potential catastrophes, which he reasonably concludes society would probably be unwilling to pay.

There are a couple of difficulties with this approach, one conceptual and one empirical. On the one hand we have estimates, however uncertain, of the benefits of controlling emissions to reach a target of, let's say, an atmospheric concentration of $CO_2$ of 450 ppm, where the benefits are just the averted damages. The natural question is then, what are the costs, to be compared to the benefits? It may be that WTP is 10% of GDP, but this is irrelevant if the costs are much lower, say 1–2% of GDP. We'll return to the question of costs, but first note another difficulty with using WTP in this context, in the event one were to seriously contemplate an empirical study.

We know from the literature on survey research that WTP is dependent on the framing of the question, and that this is especially true when people are asked to value a good (or bad) with which they have little or no experience. This result would apply with special force to a question about future consequences so far beyond the realm of current or historical experience. A second concern is that, for a credible result, in addition to a well-defined good, within the realm of experience, a credible payment mechanism needs to be part of a survey. For example, the question might be: Would you be willing to pay an additional 10% in property taxes to finance a treatment plant that would reduce the concentrations of pollutant X in a nearby lake by 90%? It's hard to see what such a mechanism would be in the setting of climate change. Of course many people have experienced natural disasters. But the potential climate catastrophe is different in that it would mean an unending and increasing series of disasters, experienced much more closely together in time and over much wider areas.

If WTP is neither an appropriate nor a reliable measure in this case, what is? Again, I suggest simply looking at the costs of controlling emissions of $CO_2$ and perhaps other GHGs, and comparing these costs to the benefits, damages averted by holding the increase in GMT below 2 °C, or some other appropriate target. There has of course been a great deal of research on this topic, which I only touch on briefly here, but the main point is that the method offers the clear advantage of being based on engineering/economic cost estimates in which we can have more confidence than estimates of WTP. There is another important advantage of the cost-of-abatement approach: we can in principle rely on the market, or market-like incentives such as a carbon tax or a cap-and-trade mechanism, to minimize the costs of attaining the objective. While academic or other publicly funded research on costs will no doubt be useful, firms will have an incentive to seek out this information and adapt it to their own circumstances. This is in clear contrast to the situation with respect to the non-climate catastrophes such as contagion or nuclear terrorism.

Now, what about the costs of meeting the target of stabilizing concentrations by the year 2100 at 450 ppm $CO_2$ or 550 ppm $CO_2$eq? Projected costs of achieving these levels appear to be relatively modest—generally in the range of 1–2% of GDP, depending on the specific target and method of computation. Drawing on an extensive set of studies, Edenhofer et al. (2010) estimate that the cost of meeting the low stabilization target of 400 ppm $CO_2$eq would come to less than 2.5% of global GDP, measured as discounted cumulative losses to the year 2100, with medium and higher stabilization levels at 450 ppm and 550 ppm approximately 1.5% and 0.8% respectively. Focusing on improvements in energy efficiency in addition to alternative energy sources, a McKinsey study puts the cost at 0.6-1.4% of global GDP in 2030 to keep the $CO_2$eq level at 500 ppm (McKinsey Global Institute 2008). The IPCC in AR4 (2007) estimates 0.6% of GDP in 2030 and 1.3% in 2050 to meet a target in the range 535–590 ppm $CO_2$eq, and in AR5 (2014) estimates 0.6% of GDP in 2030 and 1.7% in 2050 to meet a slightly more ambitious target in the range of 530–580 ppm (IPCC 2014). All of these estimates strike me as somewhat optimistic, depending on assumptions about the efficiency in reducing costs, but they are of course far below the illustrative 10% of GDP for WTP.

Alternatively to the global or macro approach, estimates on a micro level focusing on specific renewable energy sources for electricity generation reveal dramatic installation cost reductions over the past several decades, and the main sources, solar and wind, are already competitive or close in given areas (IPCC (2011)-Figure SPM.6a; Economist (2012)). Moreover, the cost of installation is expected to decrease further as adoption increases, according to Swanson's law, which suggests a 20% drop in the price of solar photovoltaic (PV) modules for every doubling of cumulative shipped volume. Wind power installation cost has also sharply declined, by more than half, from approximately \$4300/kw to less than \$2000/kw over the 1982–2008 period (Wiser and Bolinger 2009). Trends in costs of solar PV are displayed in Fig. 6.6.

Finally, in connection with any assessment of energy alternatives, we need to take into account the non-climate change-related environmental and health impacts of conventional fossil fuel use. For example, a recent study finds that air pollution from the burning of coal and oil results in \$120 billion in health costs just for the U.S. (NAS 2010). This does not account for other damages, for example water pollution from ongoing extraction and transport, and from occasional disastrous oil spills. These are real costs, though they are for the most part not reflected in the prices of oil and coal. They are however relevant in considering the net social cost of reducing GHG emissions by substituting a mix of energy conservation, increased use of renewables, continuing substitution of natural gas for coal, and perhaps construction of new nuclear plants. Alternatively, they can be considered as a benefit of low-carbon fuels, thereby reducing the net social cost of substitution. For illustrative purposes, accounting for only part of the non-climate change-related damage associated with coal-based electricity generation yields an estimate of 3.2 cents/kwh, a weighted

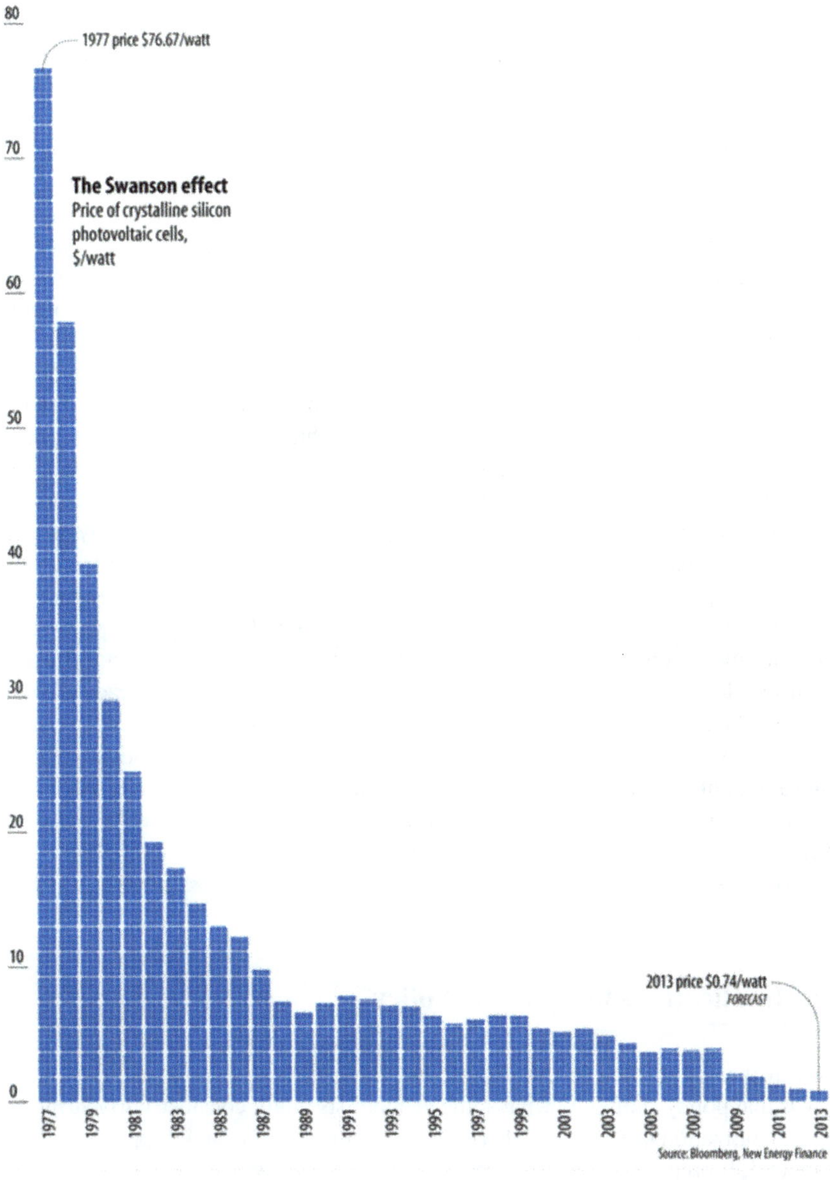


**Fig. 6.6** Trends in installation costs of solar PV power, adapted from Economist (2012)

average, where the weights are the electricity generated by each plant (NAS 2010, pp. 6–7). Subtracting this from the cost of wind-based generation would make it cheaper than coal in some places, and subtracting it from the cost of solar PV would make that source competitive in some.

There are additional differences between the other potential catastrophes and a climate-related catastrophe that make a simple adding-up of WTP, or even costs, across catastrophes not very relevant to the climate policy debate. An important one is that we can be fairly certain that, without implementing well-understood policy instruments and technologies to reduce emissions of GHGs, the kinds of consequences discussed earlier will occur. By the same token, with additional efforts at abatement, we can be fairly certain that we can at least reduce the frequency or severity of the various natural disasters. The same kind of near-certainty—of occurrence or non-occurrence related to specific policy measures—cannot be invoked with respect to the other catastrophes such as outbreak of a mega-virus or detonation of a nuclear device by a rogue state or terrorist organization.

We admittedly don't know much about the prospects for averting or reducing the probability of these catastrophes; that would be a subject for a separate research project. That said, it's not clear how much more, at what cost, can usefully be done, or that there is a connection to climate policy. Perhaps more widely deployed monitoring mechanisms for detection of outbreaks of contagious diseases, more rapid communication and treatment, and so on, would help, but given the resources already devoted to these activities any additional effort would likely involve only a tiny fraction of national budgets, much less GDP. Similarly with respect to nuclear terrorism, the U.S. and other countries are already devoting substantial resources to prevention. No doubt more could be done, but the costs would come to only a small fraction of current defense and intelligence expenditures, including those already devoted to this problem—and again, an even smaller fraction of GDP. In any event, given the relatively modest levels of additional averting expenditures (relative to GDP), controlling emissions of GHGs to hold a further increase in GMT to less than 2 °C (which would however mean an increase of more than 2 °C from the pre-industrial level) by the end of the century is not plausibly precluded by some increase in efforts to avert other catastrophes.

## 6.7 Implications for Climate Policy

What are the implications for climate policy? Based on the discussion to this point, my view is that policy should be what some economists might consider fairly stringent: target stabilization of the atmospheric concentration of $CO_2$ at 450 ppm, or $CO_2$eq at about 550 ppm. The scientific consensus is that this will reduce the probability or frequency of occurrence of some of the most damaging impacts. Moreover, it should be attainable at reasonable cost, on the order of 1–2% of world GDP. We note that some climate scientists and others now believe that even a target of 450 ppm is too high to forestall multiplying major adverse impacts, and that a long run level of something like 350 ppm is required. Hansen et al. (2013) argue that the level of cumulative emissions associated with a 2 °C warming would spur feedbacks leading to an eventual 3–4 °C warming.

An alternative approach is of course to use one or several of the established integrated assessment models (IAMs) to compute an optimal trajectory of emissions reductions, typically to the year 2100. As I noted earlier, regulations by government agencies such as EPA and OMB are at least in principle based on the social cost of carbon (SCC), in turn derived from the standard models. In addition to elucidating a link between the workings of the economy and the climate system, the models have the virtue of providing very precise solutions to the problem of how much to cut emissions and when. Another advantage is that they generate a set of shadow prices readily adapted as carbon taxes, the policy instrument favored by most economists.

The difficulty, as I also noted earlier, is that the numerical results are not reliable as a guide to policy. Before reviewing some of the reasons for this judgment I should acknowledge that, if the results were to be accepted and implemented by policymakers this would be a step in the right direction. That said, there are deficiencies that result in what Stern (2013) has characterized as gross underestimation of potential damages from climate change, especially in the long run, and in turn to less than ideal mitigation.

What are the deficiencies, and why are they significant? Functional forms and parameters of damage functions, though consistent with economic theory in a general way, are somewhat arbitrary, not grounded in empirical research even about potential impacts on the most relevant commercial sector, agriculture. Nonlinearities here and elsewhere involving extreme conditions, and irreversible, extreme and catastrophic events, are not captured. The quantitative results that form the basis for policy prescriptions are extremely sensitive to assumptions, in particular about the discount rate. Almost anything occurring in 200–300 years, no matter how catastrophic, can be rendered insignificant by a discount rate based on current rates in private capital markets. This matters because the most severe potential impacts, for example on sea level rise, will occur in that time frame and as we have argued, the appropriate discount rate in thinking about climate policy is well below market rates. Perhaps most importantly, what are likely to be the most damaging impacts are not adequately captured by the implicit assumption of perfect substitutability of conventional market goods for services of the environment that are essential to the functioning of the human economy. What would compensate for the loss of vast areas of productive land to a sea level rise of 10 m, or even 1–2 m, and the resulting mass migrations and conflicts?

An intermediate approach, taken in a number of recent studies, is to amend or augment an IAM, for example to take into account relevant nonlinearities such as extreme conditions, irreversibilities and catastrophic events. These are useful contributions, and represent a further step in the right direction, the direction the models will have to go in to provide guidance to policy. But they are clearly unable to address all of the concerns about the models. The problems, and the unknowns, are too intractable, at least at this point. It seems doubtful for example that some of the most important potential impacts, such as massive sea level rise, with all of its consequences, or major loss of ecosystem support services, can be fully delineated, much less monetized. Depending on modeling choices, an optimal increase in GMT of 2 °C, whether from pre-industrial or current levels, or for that matter a still lower level, may be indicated, and this would certainly form the basis for a policy more consistent with the scientific consensus. However, given the uncertainties it may be preferable simply to adopt the consensus of 2 °C or less as the objective of mitigation policy, on the grounds that the benefits (damages averted) will be very great, certainly much greater than the costs of achieving them, even though they cannot be fully known or monetized. Another way to think about this is, as some have suggested, to consider the costs a premium for insurance against the worst outcomes.

## 6.8 Policy Instruments

Finally, we need to consider—very briefly—how any objective can be achieved in practice. In principle, two types of instruments may be needed: a tax (or subsidy) to internalize the externality, and public investment to address the public good aspects of the problem. On the tax side, a carbon tax would clearly be the best choice, as it would give all producers and consumers an incentive to economize on fossil fuels, switch from more carbon-intensive (coal) to less (natural gas), or to alternatives (renewables). But a potential problem is that the tax would have to be very high, even to achieve, for example, the optimal trajectory in the IAMs, much less to limit further increases in GMT to 2 °C. This is illustrated in Fig. 6.7, showing carbon prices under different scenarios of the RICE model, a regionalized version of DICE (Nordhaus 2010).

What leaps out is how much higher the prices are than any taxes we are likely to see, especially going forward, for reasons that are well understood. The price per ton of carbon is $38 in 2015, rising steadily to over $400 by the end of the century (2005 prices) in the optimal run. To limit the temperature change to 2 °C, comparable prices are $79 and $904 respectively, with a spike at about $1,200 in 2085. A carbon tax can and should play a role in mitigation, but realistically will be much lower even than the level needed to achieve the RICE optimum, much less the 2 °C solution. To achieve either of these objectives, the tax will need to be supplemented.

In the absence of a tax, or to supplement one, a negative tax on low carbon energy sources, or in other words, a subsidy, in the form of tax credits to conservation and renewables, seems appropriate. But subsidies are also problematic, potentially distortionary, since they are likely to affect some activities some energy sources, and some technologies, and not others. Ideally, a subsidy should be as broad-based as possible. A good example is in fact in place in the form of the federal tax credit for residential renewable energy, which applies to solar-electric systems, solar water heating systems, fuel cells, small wind-energy systems, and geothermal heat pumps.

In addition to a tax/subsidy, another instrument that will likely be needed is public investment in fundamental, basic research in potential low-carbon energy sources. A good example is provided by a new institute at UC Berkeley, which will explore the basic science of how to capture and channel energy on the molecular or nanoscale.

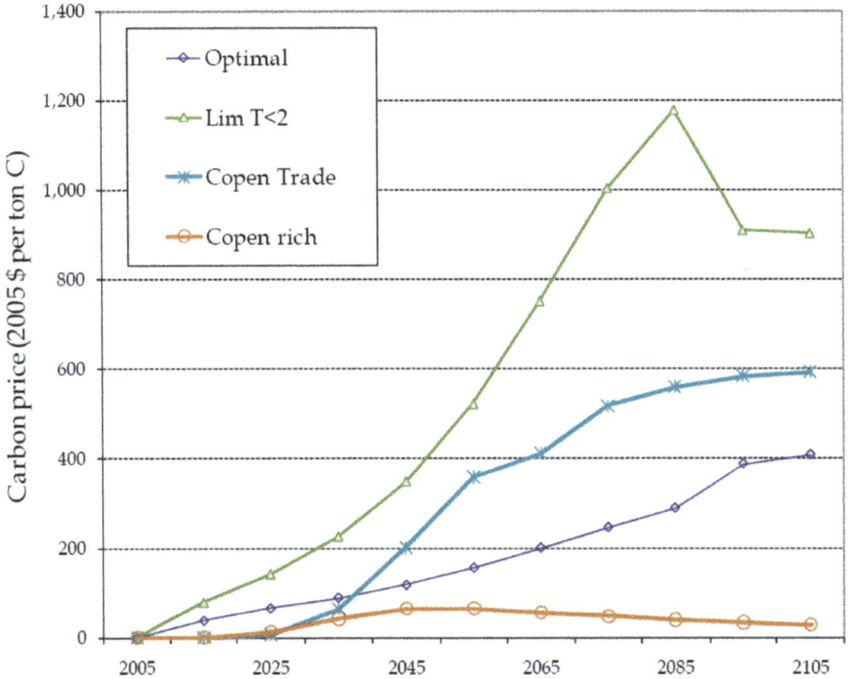


**Fig. 6.7** Carbon tax under different scenarios. *Source* Nordhaus (2010)

These are some examples, no doubt capable of expansion or improvement, of the kind of tax/subsidy/investment policies that in our judgment will be needed to achieve the desired reduction in emissions of GHGs, whether to follow the optimal trajectories in DICE and other IAMs, or to hold further increase in GMT to 2 °C or less.

# References


Ackerman, F., Stanton, E. A., & Bueno, R. (2010). Fat tails, exponents, extreme uncertainty: Simulating catastrophe in dice. *Ecological Economics*, *69*(8), 1657–1665.

Albers, H. J., Fisher, A. C., & Hanemann, W. M. (1996). Valuation and management of tropical forests: Implications of uncertainty and irreversibility. *Environmental and Resource Economics*, *8*, 39–61.

Archer, D., & Buffett, B. (2005). Time-dependent response of the global ocean clathrate reservoir to climatic and anthropogenic forcing. *Geochemistry Geophysics Geosystems*, *6*, 3.

Arrow, K. (1968). Optimal control policy with irreversible investment. *Value, Capital, and Growth. Papers in Honour of Sir John Hicks, Edinburgh University Press, Edinburgh.*

Arrow, K., Cropper, M., Gollier, C., Groom, B., Heal, G., Newell, R., et al. (2014). Should governments use a declining discount rate in project analysis? *Review of Environmental Economics and Policy*, *8*(2), 145–163.

Arrow, K., Dasgupta, P., Goulder, L., Daily, G., Ehrlich, P., Heal, G., et al. (2004). Are we consuming too much? *Journal of Economic Perspectives*, *18*(3), 147–172.

Arrow, K., & Fisher, A. (1974). Environmental preservation, uncertainty, and irreversibility. *The Quarterly Journal of Economics*, *88*(2), 312–319.

Arrow, K., & Lind, R. (1970). Uncertainty and the evaluation of public investment decisions. *The American Economic Review*, *60*(3), 364–378.

Bamber, J. L., Riva, R. E. M., Vermeersen, B. L. A., & LeBrocq, A. M. (2009). Reassessment of the potential sea-level rise from a collapse of the west antarctic ice sheet. *Science*, *324*(5929), 901–903.

Barnett, H. J., & Morse, C. (1963). *Scarcity and growth*. Baltimore: The John Hopkins Press.

Boltyanskiy, V., Gamkrelidze, R., Mishchenko, Y., & Pontryagin, L. (1962). *Mathematical theory of optimal processes*. Interscience Publishers.

Brobst, D. A. (1979). The systems approach to the analysis of resource scarcity. In V. K. Smith (Ed.), *Scarcity and growth reconsidered*. Johns Hopkins Press.

Ceronsky, M., Anthoff, D., Hepburn, C., & Tol, R. S. J. (2011). Checking the price tag on catastrophe: The social cost of carbon under non-linear climate response. Technical report, ESRI working paper, No. 392.

Cicchetti, C., & Freeman, A. (1971). Option demand and consumer surplus: Further comment. *The Quarterly Journal of Economics*, 528–539.

Cline, W. R. (1992). *The economics of global warming*. Institute for International Economics.




A. C. Fisher, *Lecture Notes on Resource and Environmental Economics*,
The Economics of Non-Market Goods and Resources 16,
https://doi.org/10.1007/978-3-030-48958-8

Cline, W. R. (2007). *Global warming and agriculture: Impact estimates by country*. Washington: Center for Global Development and Peterson Institute for International Economics.

Conrad, J. (2010). *Resource economics*. Cambridge University Press.

Conrad, J., & Clark, C. (1987). *Natural resource economics: Notes and problems*. Cambridge University Press.

Cropper, M., Aydede, S., & Portney, P. (1992). Rates of time preference for saving lives. *The American Economic Review*, *82*(2), 469–472.

Deschenes, O., & Greenstone, M. (2007). The economic impacts of climate change: Evidence from agricultural output and random fluctuations in weather. *American Economic Review*, *97*(1), 354–385.

DKRZ. (2014). Technical report. German Climate Computing Center, Laboratory for Climate Researchers.

Economist, T. (2012). Sunny uplands: Alternative energy will no longer be alternative. Technical report. Accessed on Jan 13, 2014.

Edenhofer, O., Knopf, B., Barker, T., Baumstark, L., Bellevrat, E., Chateau, B., et al. (2010). The economics of low stabilization: Model comparison of mitigation strategies and costs. *The Energy Journal*, *31*, 11–48.

Eswaran, M., Lewis, T., & Heaps, T. (1983). On the nonexistence of market equilibria in exhaustible resource markets with decreasing costs. *The Journal of Political Economy*, 154–167.

Faustmann, M. (1849). Calculation of the value which forest land and immature stands possess for forestry. English translation reprinted in *Journal of Forest Economics, 1*, 7–44 (1995).

Fisher, A. (1981). *Resource and environmental economics*. Cambridge University Press.

Fisher, A., & Karp, L. (1993). Nonconvexity, efficiency and equilibrium in exhaustible resource depletion. *Environmental and Resource Economics*, *3*(1), 97–106.

Fisher, A., Krutilla, J., & Cicchetti, C. (1972). The economics of environmental preservation: A theoretical and empirical analysis. *The American Economic Review*, *62*(4), 605–619.

Fisher, A. C., & Hanemann, M. W. (1993). Assessing climate change risks: Valuation of effects. In *Assessing surprises and nonlinearities in greenhouse warming*. Washington, D.C.: Resources for the Future.

Fisher, A. C., Hanemann, M. W., Roberts, M. J., & Schlenker, W. (2012). The economic impacts of climate change: Evidence from agricultural output and random fluctuations in weather: Comment. *American Economic Review*, *102*(7), 3749–3760.

Fisher, A. C., & Krutilla, J. V. (1975). Resource conservation, environmental preservation and the rate of discount. *Quarterly Journal of Economics*, *89*(3), 358–370.

Fisher, A. C., & Le, P. V. (2014a). Reflections on climate policy: Science, economics, and extremes. Technical report, Working paper.

Fisher, A. C., & Narain, U. (2003). Global warming, endogenous risk, and irreversibility. *Environmental and Resource Economics*, *25*(4), 395–416.

Fisher, T., & Le, P. V. (2014b). Climate policy: Science, economics, and extremes. *Review of Environmental Economics and Policy*, *8*(2), 307–327.

Hall, D., & Hall, J. (1984). Concepts and measures of natural resource scarcity with a summary of recent trends. *Journal of Environmental Economics and Management*, *11*, 363–379.

Halvorsen, R., & Smith, T. R. (1991). A test of the theory of exhaustible resources. *The Quarterly Journal of Economics*, *106*(1), 123.

Hanemann, W. M. (1989). Information and the concept of option value. *Journal of Environmental Economics and Management*, *16*(1), 23–37.

Hanley, N., Shogren, J., & White, B. (1997). *Environmental economics in theory and practice*. Palgrave Macmillan.

Hansen, J., Kharecha, P., Sato, M., Masson-Delmotte, V., Ackerman, F., et al. (2013). Assessing "dangerous climate change": Required reduction of carbon emissions to protect young people, future generations and nature. *PLoS ONE*, *8*(12)

Hartman, R. (1976). The harvesting decision when a standing forest has value. *Economic Inquiry*, *14*(March), 52–58.

Hartwick, J. (1977). Intergenerational equity and the investing of rents from exhaustible resources. *The American Economic Review, 67*(5), 972–974.

Heal, G. (2009). Climate economics: A meta-review and some suggestions for future research. *Review of Environmental Economics and Policy, 3*(1), 4–21.

Heal, G., & Park, J. (2013). Feeling the heat: Temperature, physiology & the wealth of nations. Working paper no. 19725, NBER.

Henderson, N., & Bateman, I. (1995). Empirical and public choice evidence for hyperbolic social discount rates and the implications for intergenerational discounting. *Environmental and Resource Economics, 5*(4), 413–423.

Henry, C. (1974). Investment decisions under uncertainty: the "irreversibility effect". *The American Economic Review, 64*(6), 1006–1012.

Hnyilicza, E., & Pindyck, R. (1976). Pricing policies for a two-part exhaustible resource cartel: The case of OPEC. *European Economic Review, 8*(2), 139–154.

Hope, C. (2006). The marginal impact of $co_2$ from PAGE2002: An integrated assessment model incorporating the IPCC's five reasons for concern. *Integrated Assessment Journal, 6*(1), 19–56.

Hotelling, H. (1931). The economics of exhaustible resources. *The Journal of Political Economy, 39*(2), 137–175.

Hsiang, S., Burke, M., & Miguel, E. (2013). Quantifying the influence of climate on human conflict. *Science, 341*, 1235367.

Hsiang, S., & Jina, A. (2013). Development after disaster: The multidecadal impact of tropical cyclones on long-run economic growth. Unpublished working paper.

Hyde, W. F. (1980). *Timber supply, land allocation, and economic efficiency*. Published for Resources for the Future by the Johns Hopkins University Press.

IPCC. (2011). Summary for policymakers, Technical report. IPCC Special Report on Renewable Energy Sources and Climate Change Mitigation.

IPCC. (2013). 5th annual report, summary for policymakers. Technical report, Climate change 2013: The physical science basis, Working Group I Contribution to the IPCC Fifth Assessment Report.

IPCC. (2014). Summary for policymakers, Technical report. Working Group III Contribution to the IPCC Fifth Assessment Report.

James, E. (1975). A note on uncertainty and the evaluation of public investment decisions. *The American Economic Review, 65*(1), 200–205.

Johnson, M., Bell, F., & Bennett, J. (1980). Natural resource scarcity: Empirical evidence and public policy. *Journal of Environmental Economics and Management, 7*, 256–271.

Joughin, I., Smith, B. E., & Medley, B. (2014). Marine ice sheet collapse potentially under way for the Thwaites Glacier Basin, West Antarctia. *Science, 344*(6185), 735–738.

Karp, L.(2017). Natural resources as capital. Cambridge, MA. MIT Press.

Khan, S. A., Kjær, K. H., Bevis, M., Bamber, J. L., Wahr, J., Kjeldsen, K. K., Bjørk, A. A., Korsgaard, N. J., Stearns, L. A., van den Broeke, M. R., Liu, L., Larsen, N. K., & Muresan, I. S. (2014). Sustained mass loss of the northeast Greenland ice sheet triggered by regional warming. *Nature Climate Change*.

Kneese, A. (1964). *The economics of regional water quality management* (p. 215). Baltimore: John Hopkins Press.

Kolstad, C. D. (1996). Learning and stock effects in environmental regulation: The case of greenhouse gas emissions. *Journal of Environmental Economics and Management, 31*(1), 1–18.

Kopp, R. E., Hsiang, S. M., & Oppenheimer, M. (2013, May 27–30). Empirically calibrating damage functions and considering stochasticity when integrated assessment models are used as decision tools. In *Impacts World 2013, International Conference on Climate Change Effects*, Potsdam. http://www.climate-impacts-2013.org/files/ial_kopp.pdf.

Krautkraemer, J. A. (2005). Economics of natural resource scarcity: The state of the debate. Technical report, Resources for the Future, Discussion Paper 05–14.

Krutilla, J. (1967). Conservation reconsidered. *The American Economic Review, 57*(4), 777–786.

Lemoine, D. M., & Traeger, C. P. (2012). Tipping points and ambiguity in the economics of climate change. Working Paper 18230, NBER.

Lewis, T. (1976). Monopoly exploitation of an exhaustible resource. *Journal of Environmental Economics and Management*, *3*(3), 198–204.

Malthus, T. (1798). *An essay on the principle of population: Or, A view of its past and present effects on human happiness*. London: Reeves and Turner.

Marglin, S. (1963a). The opportunity costs of public investment. *The Quarterly Journal of Economics*, 274–289.

Marglin, S. (1963b). The social rate of discount and the optimal rate of investment. *The Quarterly Journal of Economics*, 95–111.

Marshall, M. (2011). Hyperwarming climate could turn earth's poles green. *New Scientist*. Accessed on January 4, 2014.

McKinsey Global Institute. (2008). The carbon productivity challenge: Curbing climate change and sustaining economic growth. Technical report, McKinsey Climate Change Special Initiative.

Meadows, D., Meadows, D., Randers, J., et al. (1992). *Beyond the limits: Global collapse or a sustainable future*. Earthscan Publications Ltd.

Meadows, D., Meadows, D., Randers, J., & Behrens III, W. (1972). *The limits to growth: A report to the club of Rome*. New York: Universe Books.

Mendelsohn, R. (2008). Is the stern review an economic analysis? in symposium: The economics of climate change: The stern review and its critics. *Review of Environmental Economics and Policy*, *2*(1), 45–60.

Mendelsohn, R., & Dinar, A. (1999). Climate change, agriculture, and developing countries: Does adaptation matter? *The World Bank Research Observer*, *14*(2), 277–293.

Mendelsohn, R., Nordhaus, W. D., & Shaw, D. (1994). The impact of global warming on agriculture: A ricardian analysis. *American Economic Review*, *84*(4), 753–771.

Mill, J. (1848). *The principles of political economy*. Wiley Online Library.

Miller, M., & Upton, C. (1985). A test of the hotelling valuation principle. *Journal of Political Economy*, *93*(1), 1–25.

NAS. (2010). Hidden costs of energy: Unpriced consequences of energy production and use. Technical report, Committee on Health, Environmental, and other External Costs and Benefits of Energy Production and Consumption, National Research Council of the National Academies.

Nordhaus, W. (1973a). The allocation of energy resources. *Brookings Papers on Economic Activity*, *1973*(3), 529–576.

Nordhaus, W. (1974). Resources as a constraint on growth. *American Economic Review*, *64*, 22–26.

Nordhaus, W. (1992). Lethal model 2: The limits to growth revisited. *Brookings Papers on Economic Activity*, *1992*(2), 1–59.

Nordhaus, W. D. (1973b). World dynamics: Measurement without data. *The Economic Journal*, *83*(332), 1156–1183.

Nordhaus, W. D. (1993). Optimal greenhouse-gas reductions and tax policy in the "dice" model. *American Economic Review*, *83*(2), 313–317.

Nordhaus, W. D. (2007a). The challenge of global warming: Economic models and environmental policy. Technical report.

Nordhaus, W. D. (2007b, September). A review of the stern review on the economics of global warming. *Journal of Economic Literature*, *45*, 686–702.

Nordhaus, W. D. (2010). Economic aspects of global warming in a post-copenhagen environment. *Proceedings of the National Academy of Sciences (US)*.

Nordhaus, W. D., & Sztorc, P. (2013). DICE-2013R: Introduction and user's manual.

Peterson, T. C., Hoerling, M. P., Stott, P. A., & Herring, S. C. (2013). Explaining extreme events of 2012 from a climate perspective. *Special Supplement to the Bulletin of the American Meteorological Society*, *94*, 9.

Pigou, A. C. (1932). The effect of reparations on the ratio of international interchange. *The Economic Journal*, *42*(168), 532–543.

Pindyck, R. S. (2007). Uncertainty in environmental economics. *Review of Environmental Economics and Policy, 1*(1), 45–65.

Pindyck, R. S. (2013a). Climate change policy: What do the models tell us? *Journal of Economic Literature, 51*(3), 860–872.

Pindyck, R. S. (2013b). The climate policy dilemma. *Review of Environmental Economics and Policy, 7*(2), 219–237.

Poore, R. Z., Williams, S., & Tracey, C. (2011). Sea level and climate. Technical Report 002-00, USGS.

Potter, N., & Christy, F. T. (1962). *Trends in natural resource commodities: Statistics of prices, output, consumption, foreign trade and employment in the United States 1870–1957*. Baltimore: Johns Hopkins Press.

Ricardo, D. (1817). *Principles of political economy and taxation*. G. Bell and sons.

Rignot, E., Mouginot, J., Morlighem, M., Seroussi, H., & Scheuchl, B. (2014). Widespread, rapid grounding line retreat of Pine Island, Thwaites, Smith, and Kohler glaciers, West Antarctica, from 1992 to 2011. *Geophysical Research Letters, 41*, 3502–3509.

Robinson, C. (1975). The depletion of energy resources. In Pearce, D. W., & Rose, J. (Eds.), *The economics of natural resource depletion* (pp. 21–55). Wiley.

Schlenker, W., Hanemann, M. W., & Fisher, A. C. (2005). Will U.S. agriculture really benefit from global warming? accounting for irrigation in the hedonic approach. *American Economic Review, 95*(1), 395–406.

Schlenker, W., Hanemann, M. W., & Fisher, A. C. (2006). The impact of global warming on U.S. agriculture: An econometric analysis of optimal growing conditions. *The Review of Economics and Statistics, 88*(1), 113–25.

Schlenker, W., & Roberts, M. (2009). Nonlinear impact effects indicate severe damages to us crop yields under climate change. *Proceedings of the National Academy of Sciences, 106*(37), 15594–15598.

Shakhova, N., Semiletov, I., Leifer, I., Sergienko, V., Salyuk, A., Kosmach, D., Chernykh, D., Stubbs, C., Nicolsky, D., Tumskoy, V., & Gustafsson, Ö. (2013). Ebullition and storm-induced methane release from the East Siberian Arctic shelf. *Nature Geoscience*.

Shakhova, N. E., Alekseev, V. A., & Semiletov, I. P. (2010). Predicted methane emission on the East Siberian shelf. *Doklady Earth Sciences, 430*, Part 2, 190–193.

Sherwood, S. C., & Huber, M. (2010). An adaptability limit to climate change due to heat stress. *Proceedings of the National Academy of Sciences, 107*(21), 9552–9555.

Smith, V. K. (1974). *Technical change, relative prices, and environmental resource evaluation*. Baltimore: Johns Hopkins Press.

Sokolov, A. P., Stone, P. H., Forest, C. E., Prinn, R., Sarofim, M. C., Webster, M., et al. (2009). Probabilistic forecast for twenty-first-century climate based on uncertainties in emissions (without policy) and climate parameters. *Journal of Climate, 22*(19), 5175–5204.

Solow, R. (1974a). Intergenerational equity and exhaustible resources. *The Review of Economic Studies, 41*, 29–45.

Solow, R. M. (1974b). The economics of resources or the resources of economics. *The American Economic Review, 64*(2), 1–14.

Stern, N. (2007). *The economics of climate change: The Stern review*. Cambridge and New York: Cambridge University Press.

Stern, N. (2008). Richard T. Elly lecture: The economics of climate change. *American Economic Review: Papers and Proceedings, 98*(2), 1–37.

Stern, N. (2013). The structure of economic modeling of the potential impacts of climate change: Grafting gross underestimation of risk onto already narrow science models. *Journal of Economic Literature, 51*(3), 838–859.

Sterner, T., & Persson, M. U. (2008). An even sterner review: Introducing relative prices into the discounting debate. In symposium: The economics of climate change: The stern review and its critics. *Review of Environmental Economics and Policy, 2*(1), 61–76.

Stiglitz, J. (1976). Monopoly and the rate of extraction of exhaustible resources. *The American Economic Review*, *66*(4), 655–661.

Stollery, K. (1983). Mineral depletion with cost as the extraction limit: A model applied to the behavior of prices in the nickel industry. *Journal of Environmental Economics and Management*, *10*, 151–165.

Tol, R. S. J. (1997). On the optimal control of carbon dioxide emissions: An application of fund. *Environmental Modelling and Assessment*, *2*(1997), 151–63.

United States. (1952). Resources for Freedom: A Report to the President. Technical report, President's Materials Policy Commission.

Weisbrod, B. (1964). Collective-consumption services of individual-consumption goods. *The Quarterly Journal of Economics*, *78*(3), 471–477.

Weitzman, M. L. (2007, September). A review of the stern review on the economics of climate change. *Journal of Economic Literature*, *45*, 703–724.

Weitzman, M. L. (2009). On modeling and interpreting the economics of catastrophic climate change. *Review of Economics and Statistics*, *91*(1), 1–19.

Weyant, J. P. (2008). A critique of the Stern review's mitigation cost analyses and integrated assessment. In symposium: The economics of climate change: The Stern review and its critics. *Review of Environmental Economics and Policy*, *2*(1), 77–93.

Wikipedia. (2014). Wikipedia's hurricane sandy entry. Accessed on Jan 10, 2014.

Wiser, R., & Bolinger, M. (2009). 2008 wind technologies market report. Technical report, U.S., Department of Energy.

Withagen, C. (1998). Untested hypotheses in non-renewable resource economics. *Environmental and Resource Economics*, *11*(3–4), 623–634.

World Bank. (2012). Turn down the heat: Why a 4c warmer world must be avoided. Technical report, A report for the World Bank by the Potsdam Institute for Climate Impact Research and Climate Analytics.

World Commission on Environment and Development. (1987). *Our common future*. Oxford University Press.

CPSIA information can be obtained
at www.ICGtesting.com
Printed in the USA
LVHW081650110622
721065LV00003B/37

9 783030 489571